21世纪高等学校专业英语系列规划教材

土木工程

专业英语教程

主　编　黄文婧

副主编　项瑞翠

编　者（按姓氏笔画排序）

于　飞　马　波　李　妍

李玲一　张　楠　武　雪

清华大学出版社

北京交通大学出版社

· 北 京 ·

内 容 提 要

本书共由 14 个单元组成，主要内容涵盖土木工程专业前辈寄语、土木工程总论、土木工程历史、土木工程专业介绍、建筑材料、应力与应变、测量、高层建筑物、桥梁工程、道路工程、混凝土工程、钢结构工程、隧道挖掘及项目管理在土木工程中的应用等。每单元的内容包括一篇精讲课文、词汇表（含生词、词组和专有名词）、练习（含正误判断题、阅读理解题和翻译题）和一篇补充阅读材料。

本书可作为土木工程专业本科生专业英语课程的教材及其撰写学期论文和毕业论文的参考资料，也可作为土木工程技术人员和研究生提高专业英语阅读能力的参考读物。

本书封面贴有清华大学出版社防伪标签，无标签者不得销售。
版权所有，侵权必究。侵权举报电话：010 – 62782989　13501256678　13801310933

图书在版编目（CIP）数据

土木工程专业英语教程 / 黄文婧主编. —北京：北京交通大学出版社：清华大学出版社，2018.9
ISBN 978 – 7 – 5121 – 3575 – 8

Ⅰ.① 土… Ⅱ.① 黄… Ⅲ.① 土木工程 – 英语 – 高等学校 – 教材 Ⅳ.① TU

中国版本图书馆 CIP 数据核字（2018）第 132816 号

土木工程专业英语教程
TUMU GONGCHENG ZHUANYE YINGYU JIAOCHENG

责任编辑：张利军
出版发行：清 华 大 学 出 版 社　邮编：100084　电话：010 – 62776969　http：//www.tup.com.cn
　　　　　北京交通大学出版社　邮编：100044　电话：010 – 51686414　http：//www.bjtup.com.cn
印 刷 者：北京鑫海金澳胶印有限公司
经　　销：全国新华书店
开　　本：185 mm×243 mm　印张：16.25　字数：375 千字
版　　次：2018 年 9 月第 1 版　2018 年 9 月第 1 次印刷
书　　号：ISBN 978 – 7 – 5121 – 3575 – 8/TU · 178
印　　数：1 ～ 3 000 册　定价：36.00 元

本书如有质量问题，请向北京交通大学出版社质监组反映。对您的意见和批评，我们表示欢迎和感谢。
投诉电话：010 – 51686043，51686008；传真：010 – 62225406；E-mail：press@bjtu.edu.cn。

前　言

编写一本土木工程专业英语教程是编者们经过长时间的土木工程专业英语教学之后形成的构想。这本教材涵盖了土木工程专业领域的诸多重要内容，信息量大，难度适中，可以作为高等学校土木工程类专业的英语教材使用。

本书阅读材料的选择是一个较为漫长且艰苦的过程。在经过精心的选材工作之后，则进入关键的整理与编写阶段。在编写过程中，编者们经常与土木工程专业的老师及学生进行探讨，编者之间更是多次互相审阅及修正书稿，努力做到求真、求精、求新，使其更具专业性和实用性。

本书由 14 个单元组成，主要内容涵盖土木工程专业前辈寄语、土木工程总论、土木工程历史、土木工程专业介绍、建筑材料、应力与应变、测量、高层建筑物、桥梁工程、道路工程、混凝土工程、钢结构工程、隧道挖掘及项目管理在土木工程中的应用等。每单元的内容包括一篇精讲课文、词汇表（含生词、词组和专有名词）、练习（含正误判断题、阅读理解题和翻译题）和一篇补充阅读材料。

本书选材适当，专业词语阐释明晰，练习题目设计得当，有助于土木工程专业学生掌握专业英语词汇及其用法，了解专业英语的语法特点，提高专业英语阅读和翻译的能力。为了方便阅读和使用，本书课后词汇表中单词、词组及专有名词的词义详尽且编排有序。在课后练习题目的设计方面，编者们也下了一番功夫，使其尽量涵盖阅读文章的主要内容，并着重体现重点与难点，以求有效地帮助学生梳理、理解并掌握所学内容的要点。除主课文之外，每单元另附一篇饶有趣味的阅读材料供学生课外阅读，以增加学生对相关领域认知的深度和广度，提高学习的兴趣。本书既可作为土木工程专业本科生专业英语课程的教材及其撰写学期论文和毕业论文的参考资料，也可作为土木工程技术人员和研究生提高专业英语阅读能力的参考读物。

本书由黄文婧担任主编，负责全书编写大纲及编写体例的确定及统稿；项瑞翠担任副主编，协助主编确定编写大纲及编写体例并配合主编完成统稿工作。各单元具体的编写分工为：黄文婧负责编写 Unit 1 和 Unit 3，项瑞翠负责编写 Unit 2 和 Unit 4，武雪负责编写 Unit 5 和 Unit 6，李玲一负责编写 Unit 7 和 Unit 10，马波负责编写 Unit 8 和

Unit 9，于飞负责编写 Unit 11 和 Unit 12，李妍负责编写 Unit 13 和 Unit 14，张楠负责每单元补充阅读材料的搜集和整理。

  本书的出版得到了北京交通大学出版社的大力支持，许多老师和朋友也对此书的编写、审校和顺利出版给予了帮助，编者在此对他们表示衷心的感谢。

  由于编者水平有限，本书难免还存在错漏与不足之处，欢迎广大读者批评指正。

编 者
2018 年 8 月

目 录

Unit 1　On Being Your Own Engineer ………………………………………… (1)
　Section A　Text: On Being Your Own Engineer ……………………………… (1)
　Section B　Text Exploration ………………………………………………… (6)
　Section C　Supplementary Reading: Famous Engineers …………………… (11)

Unit 2　Civil Engineering ……………………………………………………… (16)
　Section A　Text: Introduction to Civil Engineering ………………………… (16)
　Section B　Text Exploration ………………………………………………… (22)
　Section C　Supplementary Reading: Construction ………………………… (28)

Unit 3　History of Civil Engineering ………………………………………… (33)
　Section A　Text: History of Civil Engineering ……………………………… (33)
　Section B　Text Exploration ………………………………………………… (41)
　Section C　Supplementary Reading: Civil Engineering in China …………… (46)

Unit 4　Civil Engineering Major ……………………………………………… (53)
　Section A　Text: Introduction to Civil and Environmental Engineering
　　　　　　　Department of College of Engineering of University of
　　　　　　　Illinois at Urbana-Champaign ………………………………… (53)
　Section B　Text Exploration ………………………………………………… (61)
　Section C　Supplementary Reading: The Department of Civil Engineering
　　　　　　　of the University of Hong Kong ………………………………… (67)

I

Unit 5　Building Materials ……………………………………………… (72)
　　Section A　Text: Building Materials ……………………………… (72)
　　Section B　Text Exploration …………………………………………… (79)
　　Section C　Supplementary Reading: Wood ……………………………… (85)

Unit 6　Stress and Strain ……………………………………………… (91)
　　Section A　Text: Stress and Strain ……………………………… (91)
　　Section B　Text Exploration …………………………………………… (97)
　　Section C　Supplementary Reading: Strength of Materials …………… (101)

Unit 7　Surveying ……………………………………………………… (109)
　　Section A　Text: Surveying ………………………………………… (109)
　　Section B　Text Exploration …………………………………………… (115)
　　Section C　Supplementary Reading: 3D Scanner ……………………… (120)

Unit 8　Tall Buildings ………………………………………………… (127)
　　Section A　Text: Tall Buildings …………………………………… (127)
　　Section B　Text Exploration …………………………………………… (134)
　　Section C　Supplementary Reading: Burj Khalifa …………………… (139)

Unit 9　Bridge Works …………………………………………………… (145)
　　Section A　Text: Bridges …………………………………………… (145)
　　Section B　Text Exploration …………………………………………… (151)
　　Section C　Supplementary Reading: Bridge Design …………………… (155)

Unit 10　Road Works …………………………………………………… (161)
　　Section A　Text: Highway Engineering ……………………………… (161)
　　Section B　Text Exploration …………………………………………… (168)
　　Section C　Supplementary Reading: Road Traffic Safety …………… (172)

Unit 11　Concrete Works ……………………………………………… (179)
　　Section A　Text: Reinforced Concrete ……………………………… (179)

Section B	Text Exploration	(186)
Section C	Supplementary Reading: Environmental Impact of Concrete	(191)

Unit 12　Steel Works (198)
Section A	Text: Structural Steel	(198)
Section B	Text Exploration	(205)
Section C	Supplementary Reading: US Steel Tower	(209)

Unit 13　Tunneling (215)
Section A	Text: Tunnel Engineering	(215)
Section B	Text Exploration	(222)
Section C	Supplementary Reading: Famous Ancient and Modern Tunnels	(227)

Unit 14　Project Management (234)
Section A	Text: Approaches and Process Groups of Project Management	(234)
Section B	Text Exploration	(242)
Section C	Supplementary Reading: Contract Conditions Used for Civil Engineering Work	(246)

Unit 1
On Being Your Own Engineer

Section A Text

On Being Your Own Engineer

The occasion for this short talk was the Civil Engineering Students Annual Awards Convocation at the University of Illinois on April 24, 1976. Parents, friends, and wives or husbands of the honor students had been invited to the Convocation. I took the occasion to speak to the wives or husbands as well as to the students who received the honors.

— Ralph B. Peck (1983)

Here at this University and in this Department that has trained so many outstanding civil engineers, you have achieved a standard of excellence that results in your recognition at this Honors Day ceremony. It gives me the greatest pleasure to congratulate you on these achievements. Here in your undergraduate career, you have become leaders in the pursuit of engineering knowledge, the first essential step in becoming a civil engineer. Excellence in undergraduate studies correlates highly with a successful engineering career in later years. I sincerely hope that the satisfaction of a successful career continues to be yours and that these honors and recognitions that you so

rightfully receive today will be only the first of many satisfactions that will come to you in your practice of civil engineering.

Yet a successful undergraduate career is not always or inevitably followed by leadership in your profession. In a changing world, in a dynamic profession such as civil engineering, how can you be sure today that you will be among the leaders of your profession 20 or 30 years from now? How can you even be sure to pick the branch of civil engineering, the particular kind of work that you will actually like the best or have the most aptitude for? Do you dare leave these matters to chance, do you dare let nature simply take its course? Nobody can predict the future and nobody can guarantee success in the future. But there are, nevertheless, many positive things you can do to shape your own career. I should like to think about some of these with you today. I believe every engineer, perhaps even while an undergraduate but certainly upon graduation, needs to form and follow his own plan for the development of his professional career. Perhaps it is an unpleasant thought, but I believe it is only realistic that nobody else is quite as interested in your career as you yourself should be. If you don't plan it yourself, it is quite possible that nobody will. On the other hand, there are too many factors, there are too many changes in a dynamic profession to permit laying out a fixed plan. The plan that you follow must be flexible and it must continually be evaluated.

To be sure, every career depends to some extent on chance, on the breaks, good or bad. But if you have followed a sound plan, you will be ready for the good breaks when they come. Those who feel they have never had favorable opportunities usually have not been ready and have not even recognized opportunities when they came.

Civil engineering projects don't exist in the classroom or in the office or in the laboratory. They exist out, in the field, in society. They are the highways, the transit systems, the landslides to be corrected, the waste disposal plants to be constructed, the bridges, the airports; they have to be built by men and machines. In my view,

Unit 1 On Being Your Own Engineer

nobody can be a good designer, a good researcher, a leader in the civil engineering profession unless he understands the methods and the problems of the builders. This understanding ought to be first hand, and if you are going to get it, you have to plan for it. Without this experience in the field, your designs may be impractical, your research may be irrelevant, or your teaching may not prepare your students properly for their profession. There are several ways in which you can get construction experience. One is by being an engineer for a builder, for a contractor. Or on the other hand, you might be an inspector for a resident engineer for the designer or owner. It doesn't matter in what capacity you work, and it doesn't take a very long time to get worthwhile experience in the field, but sometime early in your career, you should plan to get it. Since the real projects are out there in the field, you will have to go where they are to get the construction experience, and you may have to put up with a little inconvenience in order to get it.

Real problems of civil engineering design include both concept and detail. In fact, details often make or break a project. A beautifully designed cantilever bridge in Vancouver Harbor collapsed during construction because a few stiffeners were omitted on the webs of some temporary supporting beams. Spectacular failures such as this don't always follow from neglected details, but poor design, poor engineering often do. I believe every civil engineer needs a personal knowledge of the details of his branch of civil engineering. If he's going to be a geo-technical engineer, for example, he needs to know among other things exactly how borings are made and samples taken under a variety of circumstances. If he's going to be a structural engineer, he needs to know how steel structures are actually fabricated and erected. He needs to know, in other words, the state of the commercial art that plays such a large part in his profession. He needs to know how things are customarily done so that he can tell whether, for example, a commercially available sampling tool will do the job at a modest competitive price or whether some unusual tool must be developed for the particular requirements of the job. So it seems to me that you

should plan to get this sort of experience also: to spend some time on a drilling rig if you plan to be a geo-technical engineer; to work for a steel fabricator or in a design office if you intend to be a structural engineer.

How can you get this varied experience, these various components of civil engineering that are so dissimilar? I think, for the most part, you have to do it by choosing your jobs carefully and changing your job if and when it seems necessary. You may be lucky in your very first job and go to work for an organization that designs, that supervises construction, that makes its own laboratory tests, that supervises borings, and so on. If this should be true, you would be fortunate, but this is not usually the case. Even such an organization may tend to let you get stuck in one phase of their work, and you may have to persuade them from time to time to let you work in other parts of their activities. More likely you will have to change organizations, possibly even to move to another part of the country or of the world. Unfortunately you can't order the jobs that you want, when you want them, and where you want them. But you can look at every opportunity to see if it fits in your plan and to judge if the time is right to make a change. The breadth of experience is so important in a civil engineer's background that it can't be obtained any other way than by a variety of jobs or a variety of activities within a given job. You owe it to yourself and to your career to see that you get this varied background. On the other hand, while you're getting this background, you ought to avoid being a job-hopper. Each of your employers will have an investment in you. At least for a while, when you start to work for him, he will not be getting his money's worth from you. You owe him a return on his investment, you owe him good work, and you owe staying with him a reasonable minimum time while you're getting that experience.

On my first real job, I had the good fortune to be working under Karl Terzaghi. He had a good many requirements, but one of the most important was that I should keep a notebook in which I should record not just what had done

College of Engineering (ENG)

Aerospace Engineering
Agricultural & Biological Engineering
Bioengineering
Civil Engineering
Computer Engineering
Computer Science
Electrical Engineering
Engineering Mechanics
Industrial Engineering
Materials Science & Engineering
Mechanical Engineering
Nuclear, Plasma & Radiological Engineering
Physics, Engineering
Systems Engineering and Design

that day, but what had seen, and what had observed. When I went down into a tunnel heading, I should come back and sketch how the heading was being executed and how it was being braced. I soon discovered that very often, when came back, I couldn't remember exactly what had gone on in the heading. I couldn't remember exactly how the bracing fit together. In other words, my eyes had seen what was going on, but my brain didn't really register. My powers of observation were poor. But as I continued to keep this notebook, I discovered that more and more could remember what had seen, and more and more my powers of observation developed, I recommend this to you as one way to make your experience more meaningful.

An investment of ten years or so after your degree, including perhaps graduate studies as well, in accordance with a carefully planned but flexible program, will go a long way toward assuring success in your engineering career. But there is another important aspect to be considered. Any worthwhile career is demanding. It makes demands on your time and effort, and also on your family. And there are other demands on your life besides your career. Your wife or your husband will have her or his own goals and even may also have a career in mind. The demands of others in your life and the fulfillment of their goals and careers will require cooperation, adjustment, give and take. Moves from one place to another will require leaving friends, will require that your children change schools. Tensions and conflicts are inevitable and compromise and reason are necessary. You and your partner will need the best possible understanding. Many a marriage has founded on the career ambitions of one or both partners and, conversely, many a career has founded on unreasonable or non-understanding social or financial demands of the partner. There is seldom a perfect solution to this problem, but there are many good solutions. The important thing is to face up to the problems early and to keep working on them. The best engineers, I think, have achieved a reasonable balance among their goals in life. Often they can truly say that their partner in life has also been their partner in their career.

Your generation has a most exciting prospect. Don't believe for a minute the prophecies that technology has outlived its usefulness. You will have, fortunately, much more to consider than technology. You will need to be true conservationists, true ecologists in the positive sense. You will need to be involved in the social cost-benefit assessments of civil engineering work above and beyond the dollar cost-benefits. Progress in these directions will be the challenge and the great achievement of your generation, and it is an exciting prospect. But to succeed, you must be fully prepared, not poorer, but better grounded technically than your predecessors. In the next ten years, the choices you make and the experiences you get will be crucial. As Honor Students, you have taken the first necessary step with skill and distinction. All of us, your teachers, your parents, your husbands, wives, and friends wish you even greater success in the future. Indeed you must succeed, or this world will be a poorer place rather than a richer place in which to live.

Section B Text Exploration

New Words and Expressions

ceremony	['serimәuni]	n.	典礼，仪式；礼节，礼仪
component	[kәm'pәunәnt]	n.	成分，组件，要素，组成部分
compromise	['kɔmprәmaiz]	v.	妥协；危害
		n.	妥协，和解，折中
convocation	[ˌkɔnvәu'keiʃәn]	n.	召集；集会；教士会议，评议会
correlate	['kɔːrәˌleit]	v.	使有相互关系，互相有关系
customarily	['kʌstәmәrәli]	ad.	通常，习惯上
demanding	[di'mɑːndiŋ]	a.	苛求的，要求高的，吃力的
dissimilar	[di'similә]	a.	不同的
dynamic	[dai'næmik]	a.	动态的，动力的，动力学的；有活力的
ecologist	[iː'kɔlәdʒist]	n.	生态学者
erect	[i'rekt]	v.	使竖立，建造，安装
essential	[i'senʃәl]	a.	基本的，必要的；本质的；精华的
evaluate	[i'væljueit]	v.	评价，估价
fabricate	['fæbrikeit]	v.	制造，伪造，装配

Unit 1 On Being Your Own Engineer

favorable	['feivərəbl]	a.	有利的，良好的；赞成的，赞许的；讨人喜欢的
fulfillment	[ful'filmənt]	n.	履行，实行
guarantee	[ˌgærən'tiː]	n.	保证，担保；保证人；保证书；抵押品
		v.	保证，担保
inevitably	[in'evitəbli]	ad.	不可避免地，必然地
irrelevant	[i'reləvənt]	a.	不相干的，不切题的
landslide	['lændslaid]	n.	山崩，滑坡，塌方，泥石流
omit	[əu'mit]	v.	省略，遗漏，删除，疏忽
outlive	[ˌaut'liv]	v.	比……活得长，比……经久，经受住，渡过……而存在
prophecy	['prɔfisi]	n.	预言
recognition	[ˌrekəg'niʃən]	n.	识别，认出；承认，公认；重视；赞誉
rightfully	['raitfuli]	ad.	正当地，正直地
stiffener	['stifənə]	n.	加固物，加劲杆，刚性元件
supervise	['sjuːpəvaiz]	v.	监督，管理，指导
undergraduate	[ˌʌndə'grædjuət]	n.	大学生，大学肄业生
		a.	大学本科生的，大学本科生身份的

cantilever bridge	悬臂桥
cost-benefit assessment	成本收益评估
drilling rig	钻机，钻探装置
face up to	勇敢地面对
geo-technical engineer	岩土工程师
have an aptitude for	有……的才能
in pursuit of	寻求，追求
put up with	容忍
sampling tool	采样工具
steel structure	钢结构
structural engineer	结构工程师

supporting beam	支承梁
transit system	交通系统
tunnel heading	隧洞导坑
waste disposal plant	废物处理厂

University of Illinois 伊利诺伊大学
Vancouver Harbor 温哥华港

I True and false.

1. Excellence in undergraduate studies correlates highly with a successful engineering career in later years. (☐T ☐F)
2. A successful undergraduate career is always followed by leadership in the profession. (☐T ☐F)
3. Civil engineering projects exist in the classroom or in the office or in the laboratory. (☐T ☐F)
4. The plan that engineers follow must be flexible and it must continually be evaluated. (☐T ☐F)
5. The best engineers will never achieve a reasonable balance among their goals in life. (☐T ☐F)

II Choose the best answer according to the text.

1. Excellence in undergraduate studies correlates highly with _____ in later years.

Unit 1　On Being Your Own Engineer

  A. a successful engineering career   B. a rewarding job
  C. a decent social status   D. a promising further study
2. According to the author, the real projects are _____.
  A. easy to accomplish
  B. out there in the field
  C. complicated for most people
  D. beneficial to the civil engineering students
3. One should spend some time on a _____ if he/she plan to be a geo-technical engineer.
  A. steel fabricator   B. budget
  C. drilling rig   D. tunnel heading
4. _____ is so important in a civil engineer's background that it can't be obtained any other way than by a variety of jobs or a variety of activities within a given job.
  A. The salary   B. The working experience
  C. The job plan   D. The breadth of experience
5. Karl Terzaghi had a good many requirements, but one of the most important was that the author should _____ in which he should record not just what had done that day, but what had seen, and what had observed.
  A. carry a computer   B. draw a draft
  C. use a digital camera   D. keep a notebook

III　Translation.

1. 细节通常决定项目成败。一座温哥华港的设计精美的悬臂桥因临时支撑梁的梁腹板上少了几根加劲杆而在建筑过程当中坍塌了。
  A. Details often make or break a project. A beautifully designed cantilever bridge in Vancouver Harbor collapsed during construction because a few stiffeners were omitted on the webs of some temporary supporting beams.
  B. Details often decide the winning of a project. A beautifully designed cantilever bridge in Vancouver Harbor fallen apart under construction because a few stiffeners were missed on the panels of some temporary supporting beams.
  C. Details often decide the winning of a project. A beautifully designed suspended bridge in Vancouver Harbor swallowed during construction because a few stiffeners were missed on the panels of some temporary supporting beams.

D. Details often make or break a project. A well-designed suspending bridge in Vancouver Harbor will fracture during construction because a few strengthening beams were omitted on the webs of some temporary supporting beams.

2. 确信的是,在面对转机时,每个职业在一定程度上都或好或坏地取决于机会。

   A. To be sure, every profession is decided by the good or bad opportunities.
   B. Certainly, every job depends on good or bad chances, when facing the opportunities.
   C. To be sure, every career depends to some extent on chance, on the breaks, good or bad.
   D. Certainly, every career depends to some extent on advantages or disadvantages.

3. 当我下到一个隧道导坑,返回时需要对这个工程的施工状况进行概述,记录它是如何运行并支撑的。

   A. When I went down into a tunnel guidance, I should come back and sketch how the guidance was being carried and how it was being braced.
   B. When I went down into a tunnel heading, I should come back and sketch how the heading was being executed and how it was being braced.
   C. When I came into a tunnel heading, I should come back and conclude how the heading executed and how it was being supported.
   D. When I went down into a tunnel leading, I should come back and draw how the leading was being carried and how it was being sustained.

4. 许多婚姻都建立在一方或双方的职业抱负上,相反,许多人的职业生涯都建立在对配偶不合理及不可理解的社会和经济要求之上。

   A. A lot of marriages established on the career ambitions of one or both partners and, on the opposite side, a lot of marriages have founded on unrealistic or unbelievable social or financial demands of the partner.
   B. Many a marriage has founded on the professional revenge of one or both partners and, conversely, many a career has founded on unreasonable or non-understanding social or financial demands of the partner.
   C. Many a marriage has founded on the career ambitions of one or both spouses and, conversely, many a career has established on unreasonable or non-understanding partners' ridiculous requirements.
   D. Many a marriage has founded on the career ambitions of one or both partners and, conversely, many a career has founded on unreasonable or non-understanding social or financial demands of the partner.

Unit 1　On Being Your Own Engineer

5. 你的生活中别人的要求，以及他们目标和事业的实现需要合作、调整、付出和索取。
   A. The requirements of others in your life and the realization of their goals and careers will require operation, regulation, give and take.
   B. The demands of others in your life and the fulfillment of their goals and careers will require cooperation, adjustment, give and take.
   C. The supply of others in your life and the achievement of their goals and careers will require conflicts, adjustment, sacrifice and take.
   D. The requests of others in your life and the attainment of their goals and careers will require corporation, adjustment, give and take.

Section C Supplementary Reading

Famous Engineers

About Charles-Augustin de Coulomb

Charles-Augustin de Coulomb (June 14, 1736 – August 23, 1806) was a French physicist. He was best known for developing Coulomb's law, the definition of the electrostatic force of attraction and repulsion, but also did important work on friction.

Charles-Augustin de Coulomb was born in Angoulême in France. He was born in a small home near France de Revone where he was raised for 7 years before beginning his education. His parents were Henry Coulomb and Catherine Bajet. He went to school in the Collège Mazarin in Paris where his father lived. His studies included philosophy, language and literature. He also received a good education in mathematics, astronomy, chemistry and botany.

Coulomb graduated in November 1761 from écoleroyale du génie de Mézières. Over the next twenty years, he was posted to a variety of locations where he was involved in engineering. His first posting was to Brest but in February 1764 he was sent to Martinique, in the West Indies, where he was put in charge of building the new Fort Bourbon and this task occupied him until June 1772.

On his return to France, Coulomb was sent to Bouchain. However, he now began to write important works on applied mechanics and he presented his first work to the Académie des Sciences in Paris in 1773. In 1779 Coulomb was sent to Rochefort to collaborate with the Marquis de Montalembert in constructing a fort made entirely from wood near Ile d'Aix. During his period at Rochefort, Coulomb carried on his research into mechanics, in particular using the shipyards in Rochefort as laboratories for his experiments.

Upon his return to France, with the rank of Captain, he was employed at La Rochelle, the Isle of Aix and Cherbourg. He discovered first an inverse relationship of the force between electric charges and the square of its distance and then the same relationship between magnetic poles. Later these relationships were named after him as Coulomb's law.

In 1781, he was stationed at Paris. On the outbreak of the Revolution in 1789, he resigned his appointment as intendant and retired to a small estate which he possessed at Blois. He was recalled to Paris for a time in order to take part in the new determination of weights and measures, which had been decreed by the revolutionary government. He became one of the first members of the French National Institute and was appointed inspector of public instruction in 1802. His health was already very feeble and four years later he died in Paris.

Coulomb leaves a legacy as a pioneer in the field of geotechnical engineering for his contribution to retaining wall design. His name is one of the 72 names inscribed on the Eiffel Tower.

About Ralph Brazelton Peck

Dr. Ralph Brazelton Peck (June 23, 1912–February 18, 2008) was an eminent civil engineer specializing in soil mechanics. He died on February 18, 2008 from congestive

Unit 1　On Being Your Own Engineer

heart failure. He was awarded the National Medal of Science in 1975 for his development of the science and art of subsurface engineering, combining the contributions of the sciences of geology and soil mechanics with the practical art of foundation design.

Peck was born in Winnipeg, and moved to the United States at age six. In 1934 he received his Civil Engineer degree from Rensselaer Polytechnic Institute and was given a three year fellowship for graduate work in structures. On June 14, 1937 he married Marjorie Truby and obtained a Doctor of Civil Engineering degree.

After receiving his degree, he worked briefly for the American Bridge Company, then on the Chicago Subway, but Peck spent the majority of his teaching career (32 years) at the University of Illinois, initially in structures but later focused on geotechnical engineering under the influence of Karl Terzaghi, ultimately retiring in 1974. He continued to work until 2005 and was highly influential as a consulting engineer, with some 1,045 consulting projects in foundations, ore storage facilities, tunnel projects, dams, and dikes, including the Cannelton and Uniontown lock and dam construction failures on the Ohio River, the dams in the James Bay project, the Trans-Alaska Pipeline System, the Dead Sea dikes and the Rion-Antirion Bridge in Greece.

On May 8, 2008, the Norwegian Geotechnical Institute in Oslo, Norway opened the Ralph B. Peck Library. This Library is next to the Karl Terzaghi Library, also at NGI. Correspondence between these two men is part of the two working libraries. The Karl Terzaghi Library tells about the birth and growth of soil mechanics. The Ralph B. Peck Library tells about the practice of foundation engineering, and how one engineer exercised his art and science for more than sixty years. Diaries from between 1939—1941 containing Peck's work with the Chicago Subway are included along with papers and reports on many of his subsequent jobs.

During his career Peck authored over 200 publications, and served as president of the International Society of Soil Mechanics and Foundation Engineering from 1969 to 1973. He received many awards, including:

1944—The Norman Medal of the American Society of Civil Engineers;
1965—The Wellington prize of the ASCE;
1969—The Karl Terzaghi Award;
1975—The National Medal of Science, presented by President Gerald Ford;
1988—The John Fritz Medal.

In 1999, the American Society of Civil Engineers (ASCE) created the Ralph B. Peck Award to honor outstanding contributions to geotechnical engineering profession through the publication of a thoughtful, carefully researched case history or histories, or the publication of recommended practices or design methodologies based on the evaluation of case histories.

About Mao Yisheng

Mao was born in Zhenjiang, Jiangsu province. He entered Jiaotong University's Tangshan Engineering College (now Southwest Jiaotong University) and earned his bachelor's degree in civil engineering in 1916. He earned his master's degree from Cornell University and earned the first Ph. D. ever granted by the Carnegie Institute of Technology (now Carnegie Mellon University) in 1919. His doctoral treatise entitled *Secondary Stress on Frame Construction* is treasured at the Hunt Library of Carnegie Mellon University and the university constructed a statue of him on campus in his honor.

Mao was regarded as the founder of modern bridge engineering. Mao's long and productive career included designing two of the most famous modern bridges in China, the Qiantang River Bridge near Hangzhou, and the Wuhan Yangtze River Bridge in Wuhan. The Qiantang River Bridge is the first dual-purpose road-and-railway bridge designed and built by a Chinese. He also participated in the construction of China's first modern bridge—Wuhan Yangtze River Bridge. During the construction of Wuhan Yangtze River Bridge, Mao Yisheng served as chairman of the Technical Advisory Committee composed of more than 20 foreign and Chinese bridge experts, and solved 14 difficult problems relating to bridge construction. He also led the structural design of the Great Hall of the People in Beijing.

Returning to China, Mao was on the faculty of five major universities and served as president of four, such as the president and professor of Tangshan Engineering School of the Jiaotong University (now Southwest Jiaotong University), the director of Engineering

Course of National Southeastern University (later renamed National Central University and Nanjing University, and the engineering course later became Nanjing Institute of Technology and then Southeast University), the president of Engineering Course College of Beiyang University, the director of Project Office of Hangzhou Qiantang River Bridge, and the director of Bridge Planning Project Office of Transportation Ministry of Kuomintang Administration.

He significantly influenced Chinese engineering education by introducing new subject matter and innovative pedagogical approaches. In addition to his engineering expertise, he was a distinguished scholar of the history of science in China.

He advocated popular science education, and wrote *On Bridge*, *China's Arch Bridges* and many other popular science articles.

Mao served as a leader of the China Engineers Association, the Chinese Civil Engineering Society and the China Association of Science and Technology. He has also served as the president of Southwest Jiaotong University (from Tangshan Engineering College to Northern Jiaotong University to Southwest Jiaotong University), the director of the Railway Institute under the Ministry of Railway, the president of the Railway Scientific Research Center, the chairman of Beijing Science Association, the honorary president and vice-president of the China Association for Science and Technology, the vice-chairman of Jiu San Society, the vice-chairman of the Chinese People's Political Consultative Conference (CPPCC), the member of CPPCC, and the standing committee member of National People's Congress.

Mao was a senior member of International Bridge and Structural Project Association, and won the honorary title of Foreign Academician issued by the United States National Academy of Sciences.

Unit 2
Civil Engineering

Section A Text

Introduction to Civil Engineering

Civil engineering is a professional engineering discipline that deals with the design, construction, and maintenance of the physical and naturally built environment, including works like roads, bridges, buildings, canals, channels, dams, harbors, airports, irrigation projects, pipelines, power plants, and water and sewage systems. Civil engineering is traditionally broken into a number of sub-disciplines. It is the second-oldest engineering discipline after military engineering, and it is defined to distinguish non-military engineering from military engineering. Civil engineering takes place in the public sector from municipal through to national governments, and in the private sector from individual homeowners through to international companies.

The practitioners of civil engineering are named civil engineers. They plan, design, construct, and maintain infrastructures to protect the public and environmental health. The term civil engineer was established in 1750 to contrast engineers working on civil projects with the military engineers, who worked on armaments and defenses.

Over time, various sub-disciplines of civil engineering

have become recognized and much of military engineering has been absorbed by civil engineering. The broad field of civil engineering generally consists of the following technical specialties.

Structural Engineering

Structural engineering is concerned with the application of structural theory, theoretical and applied mechanics, and optimization to the design, analysis, and evaluation of building structures, bridges, cable structures, and plate and shell structures. The science of structural engineering includes the understanding of the physical properties of engineering material, the development of methods of analysis, the study of the relative merits of various types of structures and method of fabrication and construction, and the evaluation of their safety, reliability, economy, and performance.

The study of structural engineering includes such typical topics as strength of materials, structural analysis in both classical and computational methods, structural design in both steel and concrete as well as wood and masonry, solid mechanics, and probabilistic methods. The types of structures involved in a typical structural engineering work include bridges, buildings, offshore structures, containment vessels, reactor vessels, and dams. Research in structural engineering can include such topics as high-performance computing, computer graphics, computer-aided analysis and design, stress analysis, structural dynamics and earthquake engineering, structural fatigue, structural mechanics, structural models and experimental methods, structural safety and reliability, and structural stability.

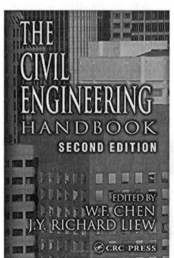

Design considerations will include strength, stiffness, and stability of the structure when subjected to loads which may be static, such as furniture or self-weight, or dynamic, such as wind, seismic, crowd or vehicle loads, or transitory, such as temporary construction loads or impact. Other considerations include cost, constructability, safety, aesthetics and sustainability.

Water Resources Engineering

This branch of civil engineering is concerned with the prediction and management of both the quality and the quantity of water in both underground (aquifers) and above ground (lakes, rivers, and streams) resources. It combines elements of hydrology, environmental science, meteorology, conservation, and resource management.

Hydraulic engineering is intimately related to the design of pipelines, water supply network, drainage facilities (including bridges, dams, channels, culverts, levees, storm sewers), and canals. Hydraulic engineers design these facilities using the concepts of fluid pressure, fluid statics, fluid dynamics, and hydraulics, among others.

Because of the diversity of the amounts of water involved and the myriad of water uses, civil engineers must deal with a multitude of water problems. Some of these problems are water supply for cities, industries, and agriculture; drainage of urban areas; and the collection of used water. Other problems deal with flows in rivers, channels, and estuaries; and flood protection; while others are concerned with oceans and lakes, hydropower generation, water transportation, etc.

Geotechnical Engineering

Geotechnical engineering studies rock and soil supporting civil engineering systems. Knowledge from the field of soil science, materials science, mechanics, and hydraulics is applied to safely and economically design foundations, retaining walls, and other structures.

Identification of soil properties presents challenges to geotechnical engineers. Boundary conditions are often well defined in other branches of civil engineering, but unlike steel or concrete, the material properties and behavior of soil are difficult to predict due to its variability and limitation on investigation. Furthermore, soil exhibits nonlinear strength, stiffness, and dilatancy, making studying soil mechanics all the more difficult.

Environmental Engineering

Environmental engineering is the contemporary term for sanitary engineering, though sanitary engineering traditionally had not included much of the hazardous waste management and environmental remediation work covered by environmental engineering. Public health engineering and environmental health engineering are other terms being used.

Environmental engineering deals with treatment of chemical, biological, or thermal wastes, purification of water and air, and remediation of contaminated sites after waste disposal or accidental contamination. Among the topics covered by environmental engineering are pollutant transport, water purification, waste water treatment, air pollution, solid waste treatment, and hazardous waste management. Environmental engineers administer pollution reduction, green engineering, and industrial ecology. Environmental engineers also compile information on environmental consequences of proposed actions.

Transportation Engineering

Transportation engineering has been one of the essential components of the civil engineering profession since its early days. It is concerned with moving people and goods efficiently, safely, and in a manner conducive to a vibrant community. This involves specifying, designing, constructing, and maintaining transportation infrastructure which includes streets, canals, highways, rail systems, airports, ports, and mass transit. It includes areas such as transportation design, transportation planning, traffic engineering, some aspects of urban engineering, queueing theory, pavement engineering, intelligent transportation system (ITS), and infrastructure management.

This subdiscipline has greatly expanded the civil engineering field to areas such as economics and financing, operations research, and management. With the rapid development of intelligent transportation systems in recent years, the transportation engineering profession has also started to make increasing use of information and communication technologies. In addition, the research and deployment of intelligent transportation systems, as well as the implementation of high-speed ground transportation systems, have gained wide attention in recent years.

Construction Engineering

Construction is the realization phase of the civil engineering process, following conception and design. It is the role of the constructor to turn the ideas of the planner and the detailed plans of the designer into physical reality.

Construction engineering involves planning and execution, transportation of materials, site development based on hydraulic, environmental, structural and geotechnical engineering. As construction firms tend to have higher business risk than other types of civil engineering firms do, construction engineers often engage in more business-like transactions, for example, drafting and reviewing contracts, evaluating logistical operations, and monitoring prices of supplies.

Municipal or Urban Engineering

Municipal engineering is concerned with municipal infrastructure. This involves specifying, designing, constructing, and maintaining streets, sidewalks, water supply networks, sewers, street lighting, municipal solid waste management and disposal, storage depots for various bulk materials used for maintenance and public works (salt, sand, etc.), public parks and cycling infrastructure. In the case of underground utility networks, it may also include the civil portion (conduits) of the local distribution networks of electrical and telecommunications services. It can also include the optimizing of waste collection and bus service networks. Some of these disciplines overlap with other civil engineering specialties, however municipal engineering focuses on the coordination of these infrastructure networks and services, as they are often built simultaneously, and managed by the same municipal authority. Municipal engineers may also design the site civil works for large buildings, industrial plants or campuses (i.e. access roads, parking lots, potable water supply, treatment or pretreatment of waste water, site drainage, etc.)

Materials Engineering

Civil engineers are involved in the design and construction of new facilities as well as the maintenance of existing structures. The decision on the choice of construction materials depends on many factors such as the cost, mechanical properties, durability, ease of construction, aesthetics, etc. The subsequent costs of operation and maintenance are also important factors to be considered in determining the economic viability of the project. Premature deterioration of the infrastructure (e. g. , roads, buildings) has serious consequences on the efficiency and profitability of other sectors of economy. Poorly constructed facilities would also affect of quality of life of their users.

To make sound decisions, engineers must be able to assess all the factors that affect the performance of a material and its interactions with the service environment. Durability is related to service life of the structure and engineers are required to optimize between cost and the duration of its intended use. The recent concern on environmental sustainability provides yet another challenge to civil engineers in their proper selection of materials for construction.

Surveying

Surveying is one of the oldest activities of the civil engineer, and remains a primary component of civil engineering. It is also one field that continues to undergo phenomenal changes due to technological developments in digital imaging and satellite positioning.

Surveying is the process by which a surveyor measures certain dimensions that occur on or near the surface of the Earth. Surveying equipment, such as levels and theodolites, are used for accurate measurement of angular deviation, horizontal, vertical and slope distances. With computerisation, electronic distance measurement (EDM), total stations, GPS surveying and laser scanning have to a large extent supplanted traditional instruments. Data collected by survey measurement is converted into a graphical representation of the Earth's surface in the form of a map. This information is then used by civil engineers, contractors and realtors to design from, build on, and trade, respectively. Although surveying is a distinct profession with separate qualifications and licensing arrangements, civil engineers are trained in the basics of surveying and mapping, as well as geographic information systems. Surveyors also lay out the routes of railways, tramway tracks, highways, roads, pipelines and streets as well as position other infrastructure, such as harbors, before construction.

In addition to the these sub-disciplines, civil engineers apply the principles of engineering management in the implementation of civil engineering projects.

Section B Text Exploration

New Words and Expressions

aesthetics [iːsˈθetiks]	n.	美学
angular [ˈæŋgjulə]	a.	有角（度）的，成角（度）的；有尖角的，尖的
armament [ˈɑːməmənt]	n.	军备，武装力量，军队，部队
aquifer [ˈækwifə]	n.	含水层，地下蓄水层
compile [kəmˈpail]	v.	编辑，编纂，编制；收集，汇集
conducive [kənˈdjuːsiv]	a.	导致……的，有助于……的
conduit [ˈkɔndit]	n.	管道，水管，导管，导线管
conservation [ˌkɔnsəˈveiʃən]	n.	保存，保藏，贮存；（对自然资源的）保护，保持
contamination [kənˌtæmiˈneiʃən]	n.	弄脏，污染，玷污；杂质，致污物，污染物
contractor [ˈkɔntræktə]	n.	立约人，订约人；承包人，承包商，承揽人
culvert [ˈkʌlvət]	n.	（横穿马路的）暗渠，涵洞，涵洞管道，电缆管道
deterioration [diˌtiəriəˈreiʃən]	n.	变坏，变质，退化，恶化，堕落
deviation [ˌdiːviˈeiʃən]	n.	脱离，偏离；偏差，偏向，离题
dilatancy [daiˈleitənsi]	n.	膨胀性，膨胀变形
dimension [diˈmenʃən; daiˈmenʃən]	n.	维度；尺寸；面积；规模；方面
discipline [ˈdisiplin]	n.	学科；纪律；训导
distinguish [disˈtiŋgwiʃ]	v.	辨别，区分，区别，分清
drainage [ˈdreinidʒ]	n.	排水，排泄；排水措施，排水系统；排水道；排出的（污）水
durability [ˌdjuərəˈbiləti]	n.	耐久性，坚固
estuary [ˈestjuəri]	n.	河口，港湾

Unit 2 Civil Engineering

fabrication	[ˌfæbri'keiʃən]	n.	制造，建造，装配，制作；构造物
hazardous	['hæzədəs]	a.	有危险的，有害的
horizontal	['hɔri'zɔntəl]	a.	水平的，横的
hydraulics	[hai'drɔːliks]	n.	水利学
hydrology	[hai'drɔlədʒi]	n.	水文学，水文地理学
infrastructure	['infrəˌstrʌktʃə]	n.	基础，基础结构，基础设施，下部结构，公共建设
intimately	['intimətli]	ad.	亲密地，密切地
irrigation	[ˌiri'geiʃən]	n.	灌溉
levee	['levi]	n.	（河的防洪）堤，堤岸，坝
logistical	[lə'dʒistikl]	a.	后勤方面的；运筹的
masonry	['meisənri]	n.	石工技术，石工工程，石工行业；石工，砖瓦工；石造（或砖砌）建筑
meteorology	[ˌmiːtiə'rɔlədʒi]	n.	气象学，气象
multitude	['mʌltiˌtjuːd]	n.	大批，许多，大量，众多
municipal	[mjuː'nisipəl]	a.	市的，都市的，市政的，市立的，市办的
myriad	['miriəd]	n.	无数，无数的人（或物）
nonlinear	[nɔn'liniə]	a.	非线性的
offshore	['ɔfʃɔː]	a.	向海的，离岸的
optimize	['ɔptimaiz]	v.	充分利用，使优化
overlap	[ˌəuvə'læp; 'əuvəlæp]	v.	与……重叠，与……部分一致
		n.	交搭，重叠
pavement	['peivmənt]	n.	人行道
phenomenal	[fi'nɔminəl]	a.	现象的，非凡的
potable	['pəutəbl]	a.	可饮的，适合饮用的
pretreatment	[priː'triːtmənt]	n.	预处理，事先处理
property	['prɔpəti]	n.	所有权，资产，财产，房地产；特性
probabilistic	[ˌprɔbəbə'listik]	a.	概率论的

reactor	[ri'æktə]	n.	反应器
realtor	['riəltə]	n.	房地产经纪人
seismic	['saizmik]	a.	地震的
sewage	['sju:idʒ]	n.	污水，阴沟污物
sewer	['sjuə]	n.	下水道，污水管，阴沟
		v.	为……铺设污水管道，用下水道排除……的污水，清洗污水管
simultaneously	[ˌsiməl'teiniəsli]	ad.	同时地
specialty	['speʃəlti]	n.	专业，专长；特制品
stiffness	['stifnis]	n.	刚度
supplant	[sə'plɑ:nt]	v.	代替，取代
telecommunications	['telikəˌmju:ni'keiʃənz]	n.	无线电通信，电信
theodolite	[θi'ɔdəlait]	n.	经纬仪
thermal	['θə:məl]	a.	热的，热量的，保热的
		n.	上升的热气流
transaction	[træn'zækʃən]	n.	办理，处理；交易；协议，协定
transitory	['trænsitəri]	a.	短暂的，暂时的，昙花一现的，瞬间即逝的
utility	[ju:'tiləti]	n.	功用，效用，实用；公用事业，公用事业设备
variability	[ˌvɛəriə'biləti]	n.	变量
vertical	['və:tikəl]	a.	垂直的，直立的
vessel	['vesəl]	n.	船，舰；（尤指盛液体的）容器，器皿
vibrant	['vaibrənt]	a.	振动的，颤动的，震颤的；充满生命力的
access road			行车通道
containment vessel			保护壳，安全壳，密闭壳
digital imaging			数字成像
distribution network			配电网

electronic distance measurement（EDM）	电子测距，光电测距
fluid pressure	流体压力，静水压力，液压系统的压力
fluid dynamics	流体动力学
fluid statics	流体静力学
GPS（global positioning system）	全球定位系统
laser scanning	激光扫描
mass transit	大容量公共运输（工具），大规模运输交通工具，（运力大的）公共交通
plate and shell structure	板壳结构
queueing theory	排队论
retaining wall	挡土墙
satellite positioning	卫星定位
solid mechanics	固体力学
storage depot	贮藏库
structural fatigue	结构疲劳
total station	全站仪（全站型电子速测仪的简称）

I True and false.

1. Civil engineering only takes place in the public sector. (□T □F)
2. Hazardous waste management and environmental remediation work are covered by environmental engineering. (□T □F)
3. Construction is the process of realizing the ideas of the planner and the detailed plans of the designer. (□T □F)
4. Transportation engineering includes the optimizing of waste collection and bus service networks. (□T □F)
5. It is absolutely necessary that civil engineers have qualifications and licenses in surveying. (□T □F)

II Choose the best answer according to the text.

1. Which of the following is not a typical topic of the study of structural engineering?
   A. Strength of materials.
   B. Solid mechanics.
   C. Structural analysis and design.
   D. Durability of materials.

2. Which of the following is not included in the water problems that civil engineers must deal with?
   A. Hydropower generation.
   B. Water purification.
   C. The collection of used waters.
   D. Water transportation.

3. Which of the following expressions about geotechnical engineering is incorrect?
   A. Geotechnical engineering studies rock and soil supporting civil engineering system.
   B. Identification of soil properties poses challenges to geotechnical engineers.
   C. Boundary conditions are often well defined in geotechnical engineering.
   D. The material properties and behavior of soil are difficult to predict.

4. The decision on the choice of construction materials depends on all the following factors except _____.
   A. air quality   B. cost and durability
   C. mechanical properties   D. ease of construction

5. Which of the following expressions about surveying is incorrect?
   A. It will continue to experience great changes due to technological developments in digital imaging and satellite positioning.
   B. It is a process by which a surveyor measures certain dimensions that occur on or near the surface of the Earth.
   C. Traditional instruments of surveying include electronic distance measurement, total stations, GPS surveying and laser scanning.
   D. Data collected by survey measurement is converted into a graphical representation of the Earth's surface in the form of a map.

Unit 2　Civil Engineering

III　Translation.

1. 土木工程是仅次于军事工程的第二古老学科，之所以将这门学科定义为土木工程，是为了把非军事工程与军事工程区分开来。
   A. Civil engineering is the second-oldest engineering discipline after military engineering, and it is defined to distinguish non-military engineering from military engineering.
   B. Civil engineering is the oldest engineering discipline after military engineering, and it is detected to separate non-military engineering and military engineering.
   C. Civil engineering is the secondly-oldest engineering discipline after military engineering, and it is defined to combine non-military engineering and military engineering.
   D. Civil engineering is the secondly-old engineering discipline after military engineering, and it is devised to set non-military engineering and military engineering apart.

2. 由于水资源类型多样且用途广泛，土木工程师必须处理水资源物理方面和管理方面的多重问题。
   A. Because of the scarcity of the amounts of water involved and the diversity of water uses, civil engineers must deal with a handful of water problems.
   B. Because of the large amount of water involved and the multiple use of water, civil engineers must deal with a handful of water problems.
   C. Because of the diversity of the amounts of water involved and the myriad of water uses, civil engineers must deal with a multiple of water problems.
   D. Because of the myriad of the amounts of water involved and the scarcity of water uses, civil engineers must deal with a variety of water problems.

3. 建筑者的作用是将设计者及其详细计划转变为有形的实物。
   A. It is the role of the constructor to become true physics the ideas of the planner and the detailed plans.
   B. It is the role of the constructor to physically realize the ideas of the planner and the detail plans.
   C. It is the role of the constructor to change the ideas of the planner and the detail plans from physically realization.
   D. It is the role of the constructor to turn the ideas of the planner and the detailed plans into physical reality.

Section C Supplementary Reading

Construction

The construction industry is one of the largest segments of business in the United States, with the percentage of the gross national product spent in construction over the last several years averaging about 10%. For 2001, the total amount spent on new construction contracts in the US is estimated at $481 billion. Of this total, about $214 billion is estimated for residential projects, $167 billion for non-residential projects, and the rest for non-building projects.

Construction is the realization phase of the civil engineering process, following conception and design. It is the role of the constructor to turn the ideas of the planner and the detailed plans of the designer into physical reality. The owner is the ultimate consumer of the product and is often the general public for civil engineering projects. Not only does the constructor have an obligation to the contractual owner, or client, but also an ethical obligation to the general public to perform the work so that the final product will serve its function economically and safely.

The construction industry is typically divided into specialty areas, with each area requiring different skills, resources, and knowledge to participate effectively in it. The area classifications typically used are residential (single- and multifamily housing), building (all buildings other than housing), heavy/highway (dams, bridges, ports, sewage-treatment plants, highways), utility (sanitary and storm drainage, water lines, electrical and telephone lines, pumping stations), and industrial (refineries, mills, power plants, chemical plants, heavy manufacturing facilities). Civil engineers can be heavily involved in all of these areas of construction, although fewer are involved in residential. Due to the differences in each of these market areas, most engineers specialize in only one or two of the areas during their careers.

Construction projects are complex and time-consuming undertakings that require the interaction and cooperation of many different persons to accomplish. All projects must be completed in accordance with specific project plans and specifications, along with other contract restrictions that may be imposed on the production operations. Essentially, all civil engineering construction projects are unique. Regardless of the similarity to other projects, there are always distinguishing elements of each project that make it unique, such as the type of soil, the exposure to weather, the human resources assigned to the project, the social and political climate, and so on. In manufacturing, raw resources are brought to a factory with a fairly controlled environment; in construction, the "factory" is set up on site, and production is accomplished in an uncertain environment.

It is this diversity among projects that makes the preparation for a civil engineering project interesting and challenging. Although it is often difficult to control the environment of the project, it is the duty of the contractor to predict the possible situations that may be encountered and to develop contingency strategies accordingly. The dilemma of this situation is that the contractor who allows for contingencies in project cost estimates will have a difficult time competing against other less competent or less cautious contractors. The failure rate in the construction industry is the highest in the US; one of the leading causes for failure is the inability to manage in such a highly competitive market and to realize a fair return on investment.

Participants in the Construction Process

There are several participants in the construction process, all with important roles in developing a successful project. The owner, either private or public, is the party that initiates the demand for the project and ultimately pays for its completion. The owner's role in the process varies considerably; however, the primary role of the owner is to effectively communicate the scope of work desired to the other parties. The designer is responsible for developing adequate working

drawings and specifications, in accordance with current design practices and codes, to communicate the product desired by the owner upon completion of the project. The prime contractor is responsible for managing the resources needed to carry out the construction

process in a manner that ensures the project will be conducted safely, within budget, and on schedule, and that it meets or exceeds the quality requirements of the plans and specifications. Subcontractors are specialty contractors who contract with the prime contractor to conduct a specific portion of the project within the overall project schedule. Suppliers are the vendors who contract to supply required materials for the project within the project specifications and schedule. The success of any project depends on the coordination of the efforts of all parties involved, hopefully to the financial advantage of all. In recent years, these relationships have become more adversarial, with much conflict and litigation, often to the detriment of the projects.

Construction Contracts

Construction projects are done under a variety of contract arrangements for each of the parties involved. They range from a single contract for a single element of the project to a single contract for the whole project, including the financing, design, construction, and operation of the facility. Typical contract types include lump sum, unit price, cost plus, and construction management.

These contract systems can be used with either the competitive bidding process or with negotiated processes. A contract system becoming more popular with owners is design-build, in which all of the responsibilities can be placed with one party for the owner to deal with. Each type of contract impacts the roles and responsibilities of each of the parties on a project. It also impacts the management functions to be carried out by the contractor on the project, especially the cost engineering function.

A major development in business relationships in the construction industry is partnering. Partnering is an approach to conducting business that confronts the economic and technological challenges in industry in the 21st century. This new approach focuses on making long-term commitments with mutual goals for all parties involved to achieve mutual success. It requires changing traditional relationships to a shared culture without regard to normal organizational boundaries. Participants seek to avoid the adversarial problems typical for many business ventures. Most of all, a relationship must be based upon trust. Although partnering in its pure form relates to a long-term business relationship for multiple projects, many single-project partnering relationships have been developed,  primarily for public owner projects. Partnering is an excellent vehicle to attain improved quality on construction projects and to avoid serious conflicts.

Partnering is not to be construed as a legal partnership with the associated joint liability. Great care should be taken to make this point clear to all parties involved in a partnering relationship.

Partnering is not a quick fix or panacea to be applied to all relationships. It requires total commitment, proper conditions, and the right chemistry between organizations for it to thrive and prosper. The relationship is based upon trust, dedication to common goals, and an understanding of each other's individual expectations and values. The partnering concept is intended to accentuate the strength of each partner and will be unable to overcome fundamental company weaknesses; in fact, weaknesses may be magnified. Expected benefits include improved efficiency and cost effectiveness, increased opportunity for innovation, and the continuous improvement of quality products and services. It can be used by either large or small businesses, and it can be used for either large or small projects. Relationships can develop among all participants in construction: owner-contractor, owner-supplier, contractor-supplier, contractor-contractor. (Contractor refers to either a design firm or a construction company.)

Goals of Project Management

Regardless of the project, most construction teams have the same performance goals:

Cost—Complete the project within the cost budget, including the budgeted costs of all change orders.

Time — Complete the project by the scheduled

completion date or within the allowance for work days.

Quality—Perform all work on the project, meeting or exceeding the project plans and specifications.

Safety — Complete the project with zero lost-time accidents.

Conflict — Resolve disputes at the lowest practical level and have zero disputes.

Project startup—Successfully start up the completed project (by the owner) with zero rework.

Basic Functions of Construction Engineering

The activities involved in the construction engineering for projects include the following basic functions:

Cost engineering — The cost estimating, cost accounting, and cost-control activities related to a project, plus the development of cost databases.

Project planning and scheduling—The development of initial project plans and schedules, project monitoring and updating, and the development of as-built project schedules.

Equipment planning and management—The selection of needed equipment for projects, productivity planning to accomplish the project with the selected equipment in the required project schedule and estimate, and the management of the equipment fleet.

Design of temporary structures — The design of temporary structures required for the construction of the project, such as concrete formwork, scaffolding, shoring, and bracing.

Contract management—The management of the activities of the project to comply with contract provisions and document contract changes and to minimize contract disputes.

Human resource management — The selection, training, and supervision of the personnel needed to complete the project work within schedule.

Project safety — The establishment of safe working practices and conditions for the project, the communication of these safety requirements to all project personnel, the maintenance of safety records, and the enforcement of these requirements.

Unit 3

History of Civil Engineering

Section A Text

History of Civil Engineering

Civil engineering is the application of physical and scientific principles for solving the problems of society, and its history is intricately linked to advances in understanding of physics and mathematics throughout history. Because civil engineering is a wide-ranging profession, including several separate specialized sub-disciplines, its history is linked to knowledge of structures, materials science, geography, geology, soils, hydrology, environment, mechanics and other fields.

Throughout ancient and medieval history most architectural design and construction was carried out by artisans, such as stonemasons and carpenters, rising to the role of master builder. Knowledge was retained in guilds and seldom supplanted by advances. Structures, roads and infrastructure that existed were repetitive, and increases in scale were incremental.

Definition

Civil engineering is the oldest branch of engineering. Not only do civil engineers design systems that interact with one another, but they are also concerned with the environment's well-being. The term "civil" was added to separate these licensed professionals from other engineers who worked on military, electrical, or mechanical projects. Architects are licensed professionals who design commercial and residential structures that are used by humans.

In ancient times, architects and builders were one and the same. Engineering was a huge part of the architect's or builder's role, especially with large construction projects such as the pyramids, the Parthenon, the Appian Way (an ancient Roman road), bridges, the aqueducts, and the Great Wall of China. Until modern times there was no clear distinction between civil engineering and architecture, and the terms were used interchangeably. However, in the 18 th century the term civil engineer began to be used to distinguish from military engineers.

Beginning

Architecture was born when people began to live in constructed dwellings and within communities. Caves were last used as habitats around 8000 BC. Early engineering was centered around food. The development of tools and ways to increase the efficiency of farming and hunting was documented, first in cave paintings and later in Egypt with hieroglyphics.

Building Materials

The history of civil engineering and architectural projects is regional in nature. The development of building design and construction on an African grassland differs greatly from building design and construction in Alaska or the mountainous regions of Peru. A major reason for differences in the development of construction techniques is the availability of local materials. Although it is possible to import materials from great distances, the historic reality is that people often used what was easily available to them when constructing buildings. This influenced architectural style and the selection of structural elements. Of course climate also has an effect on the design and construction of buildings. For example, sun-dried bricks can support loads in a dry climate but will disintegrate in wet climates.

Current day Iraq consists mostly of alluvial plains—no stone, and wood is scarce. Sun-dried bricks (clay-rich soil mixed with straw) were used as early as the fourth millennium BC. The ruins of the famed city of Babylon illustrate the use of mud-bricks.

Greece consists almost entirely of limestone and has many sources of marble. The finest source of marble is Mount Pentelicus from which the marble used in the Parthenon was cut. The pre-Columbian Inca site of Machu Picchu sits on a mountain ridge in Peru. The central buildings of Machu Picchu use the classic architectural style of dry stone walls in which blocks of stone are cut to fit together tightly without mortar.

Because of their durability, examples of stone and brick buildings from past civilizations do exist. However, few examples of ancient buildings exist where less durable materials such as wood and grass were used.

In China and in Japan, most ancient buildings were constructed of wood (except for the tile roofs). Because wood tends to deteriorate over time, few examples of ancient Chinese buildings exist. However, the Chinese have long standardized uniform features of structures through manuals and drawings that were passed down for generations. Therefore, we can determine that Chinese architecture changed little over thousands of years.

Vernacular architecture is a term used to categorize methods of construction which use locally available resources and traditions to address local needs. It is often viewed as crude and unrefined, but many modern architects have claimed inspiration from vernacular architecture. Can you think of any other examples of vernacular architecture? How about the sod houses of the great plains? The pueblos of southwestern Native Americans? Thatched huts of Hawaii?

Pyramids

The earliest large structures were the pyramids. Pyramids in Egypt were built as monuments to house the tombs of the pharaohs. The earliest known architect was Imhotep of Egypt. He was known for creating the first step pyramid at Sakkara around 2700 BC. This pyramid was a solid structure, but many early buildings were bearing wall types. Bearing walls are solid walls that provide support for each other and for the roof. The building next to the pyramid in the right image illustrates a bearing wall system. 200 years later, the Great Pyramid of Khufu was built. It

is the largest masonry structure ever built. The base measures 756 feet on each side and is 480 feet tall. The builder was Cheops, also known as Khufu. His pyramid is the only surviving of the Seven Wonders of the Ancient World.

The Egyptians were not the only ones building pyramids. In the Americas pyramids were built for religious ceremonies or scientific use. The stepped pyramid in Yucatan, Mexico was built for astronomical purposes.

Modern Pyramids

Many modern pyramids exist, including a pyramid in the courtyard of the Louvre Museum in Paris, the Luxor hotel in Las Vegas, and the Pyramid Arena in Memphis, Tennessee. Modern buildings are concerned with aesthetics as well as with functionality. Aesthetics is the quality of an object that deals with art, beauty, and taste.

The Parthenon

The Parthenon was built to house a statue of Athena, Greek goddess of war. Like other temples of its day, it was designed to be seen from the outside only. The Parthenon is an example of post-and-lintel construction: horizontal beams placed across vertical posts. This form of construction came about due to an early problem in architecture—how can door and window openings be provided in bearing walls without sacrificing support? Post-and-lintel construction is an example of an early frame system. The columns of the Parthenon are closely spaced because it is made of stone, which has little tensile capacity (unable to support wide openings). Greek temples showed a change from wide columns to slender columns. Why this movement toward slenderness and greater intermediate spans? Although the main reason was aesthetics, the desire to reach greater heights and to bridge greater spans with less material cannot be ignored.

Arches

Supporting large openings was a major problem in the early design of structures. Although very strong in compression, stone is weak in tension and cannot support the weight of the structure across a large openings. The Romans developed the arch to overcome the limitations of the post and lintel. An arch is a  curved structure for spanning an opening, designed to support a vertical load primarily by axial compression. Because they are made from smaller and lighter blocks of stone, they are easier to erect. Blocks are placed in a curved formation in such a way that they give each other support. The wedge-shaped units in the arch are called voussoir. The keystone is the voussoir at the crown of the arch, serving to lock the others in place. Can you think of some modern examples of the use of an arch? The arch is often used in the construction of bridges, tunnels, sewers, and palaces.

Vaults

The development of the arch led to the vault, which is a series of arches that form a continuous arched covering. When two of these vaults intersect, a cross vaults is created. Another name for the cross vault is the groin vault. Vaults allow for the construction of bridges, walkways, and other passages. What are some examples of vaults that you have seen? Most highway tunnels use the vault.

Domes

Another form that developed from the arch was the dome. A dome is an arrangement of several arches whose bases form a circle and the tops meet in the center. Can you think of other domes used in modern architecture?

The Pantheon is an example of an arch and dome system. It is the oldest standing domed structure in Rome. Two amazing aspects of the

Pantheon have to do with the materials used to create it. The Pantheon's concrete was a mixture of volcanic ash, lime, and a small amount of water. That mixture was packed, not poured, into place. Today we have Portland cement, a hydraulic cement that is a key ingredient in concrete and many other cementitious products such as masonry bricks and plaster. Hydraulic cement hardens by reacting with water and is water-resistant.

Colosseum

Originally known as the Flavian Amphitheatre, this arch and dome system is the largest ever built in the Roman Empire. It is considered by many to be one of the greatest works of Roman architecture and engineering. Modern sports arenas are styled after the Colosseum, so named probably due to a colossal statue of Emperor Nero which is nearby.

Water Supply

The need for a dependable water supply for domestic and agricultural purposes led to the construction of dams, canals, and aqueducts. This usually involved conducting the water over long distances. The engineering involved constructing the massive dams and solving problems associated with getting the water to flow to its destination. Since water flows downhill, a combination of canals, tunnels, and aqueducts were often required. The Roman aqueducts were built from 300 BC to 200 AD. These provided Rome with an ample supply of fresh water. In the first century AD, almost 180 million gallons of water poured daily into Rome. These aqueducts are another example of an arch and dome system.

Road System

Ancient roadways were used by the Persians and Romans for strategic and commercial purposes. Greeks needed to have roads available in the event of a religious exodus. Greek highways consisted of two wheel ruts of about 4 feet 11 inches. gauge either carved out or worn down. The roads typically had a width of 8 or 9 feet and a depth of 3 to 4 inches. Sometimes this was increased to 12 inches to make the road smoother in the rocky ground. Urban streets were mostly paved. Roman roads were created by using large blocks of stone for the base, over which broken stone or debris was spread and covered by a layer of sand and finally by large polygonal basalt blocks, with the polished top surfaces serving as the road surface. The stones were set in lime mortar. In marshy regions, the Romans used wooden causeways resting on pile foundations.

Romans were generally believed to be masters of road engineering. At the height of their power, they had constructed 50,000 miles of paved roads. No comparable road system existed outside the Roman Empire. Overland lines of communication outside the Roman Empire consisted of beaten paths. The cost of the roads was defrayed by public costs or private donations. Augustus, for example, assigned roads to wealthy Senators who were responsible for their maintenance.

Bridges

The earliest known Roman bridge, the Pons Sublicius in Rome, was made of wood and was constructed using columns and beams. The pile foundation was created by following specific steps: excavate, clear, and then drive previously charred alder, olive, or oak piles into the ground as close to each other as possible. The spaces were filled with ashes. Timber bridge is the most popular bridge during that time. Most ancient Roman bridges used the arch as the basic structure and were typically made of stone and concrete.

Although some jobs were mechanized, there was little change in building materials or methods of construction from those of the Romans until the middle of the 18th century. The same simple cranes, pulley systems, wedges, and inclined planes were still commonly used to move heavy objects. The hammer, plane, and chisel were still the tools of choice of the carpenter. However, in the middle of the 18th century, iron became cheaper and more readily available. The first iron bridge was built over the Servern in 1779. A new blast furnace nearby lowered the cost and encouraged local engineers and architects to use iron to cross the river. The use of iron allowed longer spans and lighter structures. What other materials do we have today that ancient architects and builders did not have?

Structural Steel

The Eiffel tower, as one of the last iron structures, marked the end of the iron era. In the 1890s steel replaced iron as the material of choice for large construction projects. Although it contains iron, steel also contains carbon which makes the metal harder and tougher. In addition, steel is less susceptible to corrosion. Steel frames were designed to carry the building loads so that massive load bearing walls were no longer necessary. The heights of buildings grew.

Reinforced Concrete

The concrete mixture used by the Romans was very weak in tension and bending. Experiments with improving the tensile strength of concrete by embedding metal rods into the mixture began in the mid-1800s. Eventually engineers learned how to efficiently take advantage of the combined strength of concrete in compression and steel in tension. Since that time, reinforced concrete has been used for a variety of construction projects. The flexibility of concrete allowed the use of free flowing curves and a break from the rectilinear designs of structural steel. Engineers quickly understood that reinforced concrete could be used in the design and construction of bridges. The first reinforced concrete bridge was built in France in 1907. The Cedar Avenue Bridge in Minneapolis was completed in 1929 and is considered the crowning achievement of city engineer Dristoffer Olsen Oustad, who was involved in the design of several prominent structures in the area. The bridge was added to the Register of Historic Places in 1989.

Unit 3　History of Civil Engineering

Section B　Text Exploration

New Words and Expressions

aqueduct ['ækwidʌkt]	n.	导水管；水道，水渠，沟渠；渡槽，高架渠，桥管
arch [ɑːtʃ]	n.	弓形，拱形，拱
artisan [ˌɑːtiˈzæn]	n.	工匠，技工
astronomical [ˌæstrəˈnɔmikəl]	a.	天文的
availability [əˌveiləˈbiləti]	n.	可用性，有效性，实用性
canal [kəˈnæl]	n.	运河，水道，管道，灌溉水渠
	v.	在……开凿运河
causeway [ˈkɔːzwei]	n.	堤道，砌道
cementitious [ˌsiːmenˈtiʃəs]	a.	胶结的，有黏结性的，似水泥的
chisel [ˈtʃizəl]	n.	凿子
colossal [kəˈlɔsəl]	a.	巨大的，异常的，非常的
column [ˈkɔləm]	n.	纵队，列；专栏；圆柱，柱形物
compression [kəmˈpreʃn]	n.	压力；压紧，压缩，压榨，浓缩
crane [krein]	n.	吊车，起重机；鹤
defray [diˈfrei]	v.	支出，支付
dependable [diˈpendəbl]	a.	可靠的，可信赖的，可信任的
disintegrate [disˈintigreit]	v.	使分解，使碎裂，使崩溃
dome [dəum]	n.	圆屋顶
downhill [ˈdaunˈhil]	ad.	下坡，向下
excavate [ˈekskəˌveit]	v.	挖掘，开凿
functionality [ˌfʌŋkʃəˈnæliti]	n.	功能
guild [gild]	n.	协会，行会，同业公会
habitat [ˈhæbitæt]	n.	栖息地，产地
hieroglyphic [ˌhaiərəuˈglifik]	n.	象形文字
interchangeably [ˌintəˈtʃendʒəbli]	ad.	可交换地

intersect	[ˌintə'sekt]	v.	相交，交叉
intricately	['intrəkitli]	ad.	杂乱地
keystone	['kiːstəun]	n.	主旨，基本原则；楔石
limestone	['laimstəun]	n.	石灰岩
marble	['mɑːbl]	n.	大理石
marshy	['mɑːʃi]	a.	沼泽的，湿地的
millennium	[mi'leniəm]	n.	千年期，千禧年
monument	['mɔnjumənt]	n.	纪念碑，历史遗迹
mortar	['mɔːtə]	n.	砂浆，灰浆，胶泥；结合物，黏合物
		v.	用灰泥涂抹，用灰泥结合
pharaoh	['fɛərəu]	n.	法老
pueblo	['pwebləu]	n.	印第安人的村庄
pyramid	['pirəmid]	n.	金字塔；角锥体
rectilinear	[ˌrekti'liniə]	a.	直线运动的，形成直线的，用直线围着的，由直线组成的
specialized	['speʃəlaizd]	a.	专业的，专门的
standardized	['stændədaizd]	a.	标准的，标准化的，定型的
stonemason	['stəunˌmeisən]	n.	石匠，石工
tension	['tenʃən]	n.	张力，拉力，膨胀力
tomb	[tuːm]	n.	坟墓
uniform	['juːnifɔːm]	a.	统一的，一致的
unrefined	[ˌʌnri'faind]	a.	未提炼的
vault	[vɔːlt]	n.	拱顶
voussoir	[vuː'swɑː]	n.	拱石
water-resistant	['wɔːtəriˌzistənt]	a.	抗水的
wedge-shaped	['wedʒʃeipt]	a.	楔形的
well-being	['wel'biːŋ]	n.	幸福，康乐
alluvial plains			冲积平原
axial compression			轴向挤压
be susceptible to			对……敏感，易患……，易受……影响

bearing wall	承重墙
cross vault	交叉穹窿
dry stone wall	干砌石墙
horizontal beam	水平梁
hydraulic cement	水凝水泥，水硬水泥
masonry structure	砌体结构
pile foundation	桩基
polygonal basalt blocks	多边形玄武岩块
Portland cement	波特兰水泥，硅酸盐水泥，普通水泥
post-and-lintel construction	连梁柱结构
pulley system	滑轮系统
sod house	草泥墙房屋
solid structure	立体结构
solid wall	实体墙
stepped pyramid	阶梯金字塔
tensile capacity	抗拉强度
thatched hut	茅草屋
tile roof	瓦屋顶
vertical load	垂直荷载
vertical post	立柱
volcanic ash	火山灰
wheel rut	车辙

Proper Names

Alaska	（美国）阿拉斯加州
Appian Way	亚壁古道
Babylon	巴比伦
Colosseum	罗马圆形大剧场

Iraq	伊拉克
Louvre	（法）罗浮宫
Machu Picchu	马丘比丘（古城，位于秘鲁中部偏南）
Minneapolis	明尼阿波里斯（美国的一座城市）
Parthenon	帕台农神殿
Pantheon	万神殿
Pentelicus	彭忒利科斯山
Peru	秘鲁
Sakkara	萨卡拉（埃及北部的村庄）
the Great Pyramid of Khufu	胡夫金字塔

I True and false.

1. The Great Pyramid of Khufu is the largest masonry structure ever built. (□T □F)
2. Post-and-lintel construction is the solution to the problem of providing door and window openings in bearing walls without sacrificing support. (□T □F)
3. Modern sports arenas are styles after the Pantheon. (□T □F)
4. Civil engineering projects exist in the classroom or in the office or in the laboratory. (□T □F)
5. The plan that engineers follow must be flexible and it must continually be evaluated. (□T □F)

II Choose the best answer according to the text.

1. What is drawn in the picture?

Unit 3 History of Civil Engineering

    A. Dome.   B. Vault.   C. Groin vault.   D. Keystone.

2. Stone is very strong in _____ and weak in _____ .

    A. compression, tension   B. tension, compression

    C. volume, size   D. size, volume

3. What system is shown in the picture?

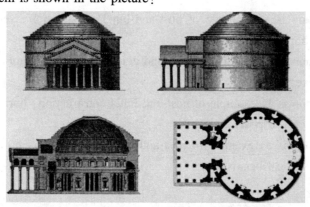

    A. Vernacular system.

    B. Arch and dome system.

    C. Masonry system.

    D. Timber system.

4. The Eiffel Tower marked _____ .

    A. the beginning of the iron era

    B. the falling of the steel era

    C. the beginning of the steel era

    D. the end of the iron era

5. Because _____ , few examples of ancient Chinese buildings exist.

    A. stone is weak in tension

    B. wood is strong in compression

    C. so many drawings were kept well

    D. wood tends to deteriorate over time

III Translation.

1. 今天的伊拉克绝大部分处于冲积平原，没有石头，木材也很少见。

    A. Current day Iraq consists mostly of alluvial plains—no stone, and wood is scarce.

B. Current day Iran consists mostly of basins—no stone, and wood is scarce.

C. Current day Iraq consists mostly of plateaus—no stone, and wood is scarce.

D. Current day Iraq consists mostly of deltas—no stone, and wood is scarce.

2. 帕台农神殿是一个典型的梁连柱结构的例子：水平的横梁搭建在垂直的立柱之上。

A. Parthenon is an example of post-and-lintel construction: vertical beams placed across horizontal posts.

B. Parthenon is an example of arch and dome construction: horizontal beams placed across vertical posts.

C. Parthenon is an example of post-and-lintel construction: horizontal beams placed across vertical posts.

D. Parthenon is an example of post-and-lintel construction: horizontal posts placed across vertical beams.

Section C Supplementary Reading

Civil Engineering in China

The Forbidden City

The Forbidden City was the Chinese imperial palace from the Ming Dynasty to the end of the Qing Dynasty—the years 1420 to 1912. It is located in the center of Beijing, China, and now houses the Palace Museum. It served as the home of emperors and their households as well as the ceremonial and political center of Chinese government for almost 500 years.

Constructed from 1406 to 1420, the complex consists of 980 buildings and covers 72 ha (over 180 acres). The palace complex exemplifies traditional Chinese palatial architecture, and has influenced cultural and architectural developments in East Asia and elsewhere. The Forbidden City was declared a World Heritage Site in 1987, and is listed by UNESCO as the largest collection of preserved ancient wooden structures in the world. Since 1925 the Forbidden City has been under the charge of the Palace Museum, whose extensive collection of artwork were built upon the imperial collections of the Ming and Qing dynasties. Part of the museum's former collection is now located in the National Palace Museum in Taipei. With over 14.6 million annual visitors, the Palace Museum is the most visited art museum in the world.

The Forbidden City is surrounded by a 7.9 meters (26 ft) high city wall and a 6 meters (20 ft) deep by 52 meters (171 ft) wide moat. The walls are 8.62 meters (28.3 ft) wide at the base, tapering to 6.66 meters (21.9 ft) at the top. These walls served as both defensive walls and retaining walls for the palace. They were constructed with a rammed earth core, and surfaced with three layers of specially baked bricks on both sides, with the interstices filled with mortar. At the four corners of the wall sit towers with intricate roofs boasting 72 ridges, reproducing the Pavilion of Prince Teng and the Yellow Crane Pavilion as they appeared in Song Dynasty paintings. These towers are the most visible parts of the palace to commoners outside the walls, and much folklore is attached to them. According to one legend, artisans could not put a corner tower back together after it was dismantled for renovations in the early Qing Dynasty, and it was only rebuilt after the intervention of carpenter-immortal Lu Ban.

The wall is pierced by a gate on each side. At the southern end is the main Meridian Gate. To the north is the Gate of Divine Might, which faces Jingshan Park. The east and west gates are called the "East Glorious Gate" and "West Glorious Gate". All gates in the Forbidden City are decorated with a nine-by-nine array of golden door nails, except for the

East Glorious Gate, which has only eight rows. The Meridian Gate has two protruding wings forming three sides of a square (Wumen, or Meridian Gate, Square) before it. The gate has five gateways. The central gateway is part of the Imperial Way, a stone flagged path that forms the central axis of the Forbidden City and the ancient city of Beijing itself, and leads all the way from the Gate of China in the south to

Jingshan in the north. Only the Emperor may walk or ride on the Imperial Way, except for the Empress on the occasion of her wedding, and successful students after the Imperial Examination.

Anji Bridge

The bridge is also commonly known as the Zhaozhou Bridge, after Zhao County, which was formerly known as Zhaozhou. Another name for the bridge is the Great Stone Bridge. It crosses the Xiao River in Zhao County, approximately 40 kilometers southeast of the provincial capital Shijiazhuang. It is a pedestrian bridge and is currently open to the public.

The Anji Bridge is about 50 meters (160 ft) long with a central span of 37.37 meters (122.6 ft). It stands 7.3 meters (24 ft) tall and has a width of 9 meters (30 ft). The arch covers a circular segment less than half of a semicircle (84°) and with a radius of 27.27 meters (89.5 ft), has a rise-to-span ratio of approximately 0.197 (7.3 to 37 meters). This is considerably smaller than the rise-to-span ratio of 0.5 of a semicircular arch bridge and slightly smaller than the rise-to-span ratio of 0.207 of a quarter circle. The arch length to span ratio is 1.1, less than the arch-to-span ratio of 1.57 of a semicircle arch bridge by 43%, thus the saving in material is about 40%, making the bridge lighter in weight. The elevation of the arch is about 45°, which subjects the abutments of the bridge to downward force and sideways force. This bridge was built in 605.

The central arch is made of 28 thin, curved limestone slabs which are joined with iron dovetails. This allows the arch to adjust to shifts in its supports and prevents the bridge from collapsing even when a segment of the arch breaks. The bridge has two small side arches on either side of the main arch. These side arches serve two important functions: first, they reduce the total weight of the bridge by about 15.3% or approximately 700 tons, which is vital because of the low rise-to-span ratio and the large forces on the abutments it creates. Second, when the bridge is submerged

during a flood, they allow water to pass through, thereby reducing the forces on the structure of the bridge.

Li Chun's innovative spandrel-arch construction, while economizing in materials, was also of considerable aesthetic merit. An inscription left on the bridge by the Tang Dynasty officials seventy years after its construction reads:

"This stone bridge over the Xiao River is the result of the work of the Sui engineer Li Chun. Its construction is indeed unusual, and no one knows on what principle he made it. But let us observe his marvelous use of stone-work. Its convexity is so smooth, and the wedge-shaped stones fit together so perfectly. How lofty is the flying-arch! How large is the opening, yet without piers! Precise indeed are the cross-bonding and joints between the stones, masonry blocks delicately interlocking like mill wheels, or like the walls of wells; a hundred forms (organized into) one. And besides the mortar in the crevices there are slender-waisted iron cramps to bind the stones together. The four small arches inserted, on either side two, break the anger of the roaring floods, and protect the bridge mightily. Such a master-work could never have been achieved if this man had not applied his genius to the building of a work which would last for centuries to come."

Dujiangyan

The Dujiangyan is an ancient irrigation system in Dujiangyan City, Sichuan, China. Originally constructed around 256 BC by the State of Qin as an irrigation and flood control project, it is still in use today. The system's infrastructure is on the Min River (Minjiang), the longest tributary of the Yangtze. The area is in the west part of the Chengdu Plain, at the

confluence between the Sichuan basin and the Tibetan plateau. Originally the Min River rushed down from the Min Mountains, but slowed abruptly after reaching the Chengdu Plain, filling the watercourse with silt, which made the nearby areas extremely prone to floods. Li Bing, then governor of Shu for the state of Qin, and his son headed the construction of the Dujiangyan, which harnessed the river using a new method of channeling and dividing the water rather than simply following the old way of dam building. It is still in use today to irrigate over 5 300 square kilometers (2,000 square miles) of land in the region. The Dujiangyan, the Zhengguo Canal in Shaanxi and the Lingqu Canal in Guangxi are collectively known as the "three great hydraulic engineering projects of the Qin Dynasty".

Irrigation Head

Li Bing's Irrigation System consists of three main constructions that work in harmony with one another to ensure against flooding and keep the fields well supplied with water:

The Yuzui or Fish Mouth Levee, named for its conical head that is said to resemble the mouth of a fish, is the key part of Li Bing's construction. It is an artificial levee that divides the water into inner and outer streams. The inner stream is deep and narrow, while the outer stream is relatively shallow but wide. This special structure ensures that the inner stream carries approximately 60% of the river's flow into the irrigation system during dry season. While during flood, this amount decreases to 40% to protect the people from flooding. The outer stream drains away the rest, flushing out much of the silt and sediment.

The Feishayan or Flying Sand Weir has a 200 meter-wide opening that connects the inner and outer streams. This ensures against flooding by allowing the natural swirling flow of the water to drain out excess water from the inner to the outer stream. The swirl also drains out silt and sediment that failed to go into the outer stream. A modern reinforced concrete weir has replaced Li Bing's original weighted bamboo baskets.

The Baopingkou or Bottle-Neck Channel, which Li Bing gouged through the mountain, is the final part of the system. The channel distributes the water to the farmlands in the Chengdu Plain, whilst the narrow entrance that gives it its name, works as a check gate, creating the whirlpool flow that carries away the excess water over Flying Sand Fence, to ensure against flooding.

Anlan Suspension Bridge

Anlan or Couple's Bridge spans the full width of the river connecting the artificial island to both banks and is known as one of the Five Ancient Bridges of China. Li Bing's original

Zhupu Bridge only spanned the inner stream connecting the levee to the foot of Mount Yulei. This was replaced in the Song Dynasty by Pingshi Bridge which burned down during the wars that marked the end of the Ming Dynasty.

In 1803 during the Qing Dynasty a local man named He Xiande and his wife proposed the construction of a replacement, made of wooden plates and bamboo handrails, to span both streams and this was nicknamed Couple's Bridge in their honor. This was demolished in the 1970s and replaced by a modern bridge.

Great Wall

The Great Wall of China is a series of fortifications made of stone, brick, rammed earth, wood, and other materials, generally built along an east-to-west line across the historical northern borders of China to protect the Chinese states and empires against the raids and invasions of the various nomadic groups of the Eurasian Steppe. Several walls were being built as early as the 7th century BCE; these, later joined together and made bigger and stronger, are now

collectively referred to as the Great Wall. Especially famous is the wall built 220 – 206 BCE by Qin Shi Huang, the first Emperor of China. Little of that wall remains. Since then, the Great Wall has been rebuilt, maintained, and enhanced; the majority of the existing wall is from the Ming Dynasty (1368 – 1644).

Other purposes of the Great Wall have included border controls, allowing the imposition of duties on goods transported along the Silk Road, regulation or encouragement of trade and the control of immigration and emigration. Furthermore, the defensive characteristics of the Great Wall were enhanced by the construction of watch towers, troop barracks, garrison stations, signaling capabilities through the means of smoke or fire, and the fact that the path of the Great Wall also served as a transportation corridor.

The Great Wall stretches from Dandong in the east to Lop Lake in the west, along an arc that roughly delineates the southern edge of Inner Mongolia. A comprehensive archaeological survey, using advanced technologies, has concluded that the Ming walls measure 8,850 km (5,500 mi). This is made up of 6,259 km (3,889 mi) sections of actual wall, 359 km (223 mi) of trenches and 2,232 km (1,387 mi) of natural defensive barriers such as hills and rivers. Another archaeological survey found that the entire wall with all of its branches measure out to be 21,196 km (13,171 mi).

Unit 4
Civil Engineering Major

Section A Text

Introduction to Civil and Environmental Engineering Department of College of Engineering of University of Illinois at Urbana-Champaign

Educational Objectives

The College of Engineering prepares men and women for professional careers in engineering and related positions in industry, commerce, education, and government. Graduates at the bachelor's level are prepared to begin the practice of engineering or to continue their formal education at a graduate school of their choice. Based on the mission and vision statement of the college, each engineering program has developed educational objectives which are broad statements that describe what graduates are expected to attain within a few years of graduation. In general, all the programs provide students with a comprehensive education that includes in-depth  instruction in their chosen fields of study. The programs are designed to emphasize analysis and problem solving and to provide exposure to open-ended problems and design methods. The courses are taught in a manner that fosters teamwork, communication skills, and individual professionalism, including ethics and environmental awareness. The classroom experiences, along with outside activities, prepare students for lifetimes of continued learning and leadership. Thus, the engineering programs enable graduates to make significant contributions in their chosen fields while at the same time recognizing their responsibilities to society.

Outcomes and Assessment

To accomplish the educational objectives and to fulfill current engineering accreditation criteria, all engineering programs provide the knowledge, experience, and opportunities necessary for students to demonstrate their attainment of the following outcomes:

(1) An ability to apply knowledge of mathematics, science, and engineering.

(2) An ability to design and conduct experiments, as well as to analyze and interpret data.

(3) An ability to design a system, component, or process to meet desired needs within realistic constraints such as economic, environmental, social, political, ethical, health and safety, manufacturability, and sustainability.

(4) An ability to function on multidisciplinary teams.

(5) An ability to identify, formulate, and solve engineering problems.

(6) An understanding of professional and ethical responsibility.

(7) An ability to communicate effectively.

(8) The broad education necessary to understand the impact of engineering solutions in a global, economic, environmental, and societal context.

(9) A recognition of the need for, and an ability to engage in lifelong learning.

(10) A knowledge of contemporary issues.

(11) An ability to use the techniques, skills, and modern engineering tools necessary for engineering practice.

Professional Component

Each engineering program also contains a professional component, as required for accreditation, that is consistent with the objectives of the program and the institution. The professional component includes:

(1) One year of a combination of college-level mathematics and basic sciences (some with experimental experience) appropriate to the discipline. Basic sciences are defined as biological, chemical, and physical sciences.

(2) One and one-half years of engineering topics, consisting of engineering sciences and engineering design appropriate to the student's field of study.

(3) A general education component that complements the technical content of the program and is consistent with the objectives of the program and the institution.

Students in engineering programs are prepared for engineering practice through a curriculum culminating in a major design experience based on the knowledge and skills acquired in earlier course work and incorporating appropriate engineering standards and multiple realistic constraints.

Civil and Environmental Engineering Department

Civil engineering is a profession that applies the basic principles of science in conjunction with mathematical and computational tools to solve problems associated with developing and sustaining civilized life on our planet. Civil engineering works are generally one-of-a-kind projects; they are often grand in scale; and they usually require cooperation among professionals of many different disciplines. The completion of a civil engineering project involves the solution of technical problems in which uncertainty of information and myriad non-technical factors often play a significant role. Some of the most common examples of civil engineering works include bridges, buildings, dams, airports, highways, tunnels, and water distribution systems. Civil engineers are concerned with flood control, landslides, air and water pollution, and the design of facilities to withstand earthquakes and other natural hazards, in addition to protecting our environment for a sustainable future.

The civil engineering program comprises seven main areas (construction engineering and management, construction materials engineering, environmental engineering, geotechnical engineering, environmental hydrology and hydraulics, structural engineering, and transportation engineering) and three cross-cutting programs (sustainable and resilient infrastructure systems; energy, water, and environmental sustainability; and societal risk management). Although each

area has its own special body of knowledge and engineering tools, they all rely on the same fundamental core principles. Civil engineering projects often draw expertise from many of these areas and programs.

The civil and environmental engineering program's education objectives are to educate students to:

(1) Successfully enter the civil and environmental engineering profession as practicing engineers and consultants with prominent companies and organizations in diverse areas that include structural, transportation, geotechnical, materials, environmental, and hydrologic engineering; construction management; or other related or emerging fields.

(2) Pursue graduate education and research at major research universities in civil and environmental engineering, and related fields.

(3) Pursue professional licensure.

(4) Advance to leadership positions in the profession.

(5) Engage in continued learning through professional development.

(6) Participate in and contribute to professional societies and community services.

Program Review and Approval

To qualify for the degree of Bachelor of Science in Civil Engineering, each student's academic program plan must be reviewed by a standing committee of the faculty (the Program Review Committee) and approved by the Associate Head of Civil and Environmental Engineering in charge of undergraduate programs. This review and approval process ensures that individual programs satisfy the educational objectives and all of the requirements of the civil engineering program, that those programs do not abuse the substantial degree of flexibility that is present in the curriculum, and that the career interests of each student are cultivated and served.

Unit 4 Civil Engineering Major

Overview of Curricular Requirements

The curriculum requires 128 hours for graduation and is organized as follows:

1. Orientation and professional development

These courses introduce the opportunities and resources your college, department, and curriculum can offer you as you work to achieve your career goals. They also provide the skills to work effectively and successfully in the engineering profession.

Course List (Total Hours: 1)
CEE 195: About Civil Engineering (1 hour)
CEE 495: Professional Practice (0 hour)
ENG 100: Engineering Orientation (0 hour)

2. Foundational mathematics and science

These courses stress the basic mathematical and scientific principles upon which the engineering discipline is based.

Course List (Total Hours: 34)
CHEM 102: General Chemistry I (3 hours)
CHEM 103: General Chemistry Lab I (1 hour)
CHEM 104: General Chemistry II (3 hours)
CHEM 105: General Chemistry Lab II (1 hour)
MATH 221: Calculus I (4 hours)
MATH 225: Introductory Matrix Theory (2 hours)
MATH 231: Calculus II (3 hours)
MATH 241: Calculus III (4 hours)
MATH 285: Introductory Differential Equations (3 hours)
PHYS 211: University Physics: Mechanics (4 hours)
PHYS 212: University Physics: Electricity & Magnetics (4 hours)
PHYS 213: University Physics: Thermal Physics (2 hours)

3. Civil engineering technical core

These courses stress fundamental concepts and basic laboratory techniques that comprise the common intellectual understanding of civil engineering.

Course List (Total Hours: 25)
CEE 201: Systems Engineering & Economics (3 hours)
CEE 202: Engineering Risk & Uncertainty (3 hours)

CS 101: Introductory Computing: Engineering & Science (3 hours)
GE 101: Engineering Graphics & Design (3 hours)
TAM 211: Statics (3 hours)
TAM 212: Introductory Dynamics (3 hours)
TAM 251: Introductory Solid Mechanics (3 hours)
TAM 335: Introductory Fluid Mechanics (4 hours)

4. Science elective

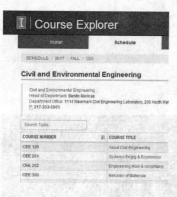

This elective allows the student to gain additional depth in science. The course should be selected according to the requirements and recommendations for the selected area of study, which is subject to approval by the faculty Program Review Committee.

Course List (Total Hours: 3)

Science elective, selected in accord with recommendations for the chosen primary field in civil engineering as outlined in the Civil Engineering Undergraduate Handbook.

5. Civil engineering technical electives

This course work is designed to give each student a broad background in the areas of civil engineering through the core courses and to allow each student to develop a focused program through advanced technical electives in chosen primary and secondary fields. There are seven areas of study which include:

Construction Engineering and Management
Construction Materials Engineering
Environmental Engineering
Environmental Hydrology and Hydraulic Engineering
Geotechnical Engineering
Structural Engineering
Transportation Engineering

In addition to the areas of study, three cross-cutting programs can be chosen by students. They include:

Sustainable and Resilient Infrastructure Systems
Energy-Water-Environment Sustainability
Societal Risk Management

The fundamental principles of civil engineering design and the behavior of civil engineering systems are emphasized throughout the course work. The specific choices of courses in this category are made through the submission of the Plan of Study, which is subject to approval by the faculty Program Review Committee.

At least 34 hours of civil engineering technical courses should be elected, including 15-16 hours of civil engineering core courses, 12-13 hours of primary field advanced technical electives and 6 hours of secondary field advanced technical electives.

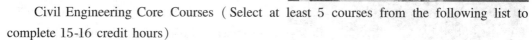

Course List (Total Hours: 34)

Civil Engineering Core Courses (Select at least 5 courses from the following list to complete 15-16 credit hours)

CEE 300: Behavior of Materials (4 hours)
CEE 310: Transportation Engineering (3 hours)
CEE 320: Construction Engineering (3 hours)
CEE 330: Environmental Engineering (3 hours)
CEE 350: Water Resources Engineering (3 hours)
CEE 360: Structural Engineering (3 hours)
CEE 380: Geotechnical Engineering (3 hours)

Primary Field Advanced Technical Electives (12-13 hours)

Select courses from approved lists for appropriate programs of study within the seven areas or three interdisciplinary programs of civil engineering. Design experience is distributed in 200-level, 300-level, and 400-level CEE courses including integrated design courses.

Secondary Field Advanced Technical Electives (6 hours)

Select courses from approved lists to complement the primary area and add breadth to the program of study.

6. Liberal education

The liberal education courses develop students' understanding of human culture and society, build skills of inquiry and critical thinking, and lay a foundation for civic engagement and lifelong learning.

Course List (Total Hours: 18)

ECON 10: Microeconomic Principles (Recommended) or ECON 103: Macroeconomic Principles (3 hours)

Electives from the campus General Education social & behavioral sciences list (3 hours)

Electives from the campus General Education humanities & the arts list (6 hours)

Electives either from a list approved by the college or from the campus General Education lists for social & behavioral sciences or humanities & the arts (6 hours)

Students must also complete the campus cultural studies requirement by completing one western/comparative culture(s) course and one non-western/US minority culture(s) course from the General Education cultural studies lists. Most students select liberal education courses that simultaneously satisfy these cultural studies requirements. Courses from the western and non-western lists that fall into free electives or other categories may also be used satisfy the cultural studies requirements.

7. Composition

These courses teach fundamentals of expository writing.

Course List (Total Hours: 7)

RHET 105: Writing and Research (4 hours)

BTW 261: Principles Technical Communication (satisfies the Advanced Composition requirement) (3 hours)

8. Free electives

These unrestricted electives, subject to certain exceptions as noted at the College of Engineering advising website, give the student the opportunity to explore any intellectual area of unique interest. This freedom plays a critical role in helping students to define research specialties or to complete minors.

Course List (Total Hours: 6)

Free electives. Additional unrestricted course work, subject to certain exceptions as noted at the College of Engineering advising Web site, so that there are at least 128 credit hours earned toward the degree. (6 hours)

Unit 4 Civil Engineering Major

Section B Text Exploration

New Words and Expressions

abuse [əˈbjuːz]	v.	滥用，妄用，误用；虐待，辱骂
accreditation [əˌkrediˈteiʃən]	n.	委托，委派；委任状；水准鉴定（或评估，鉴定合格，认可）
appropriate [əˈprəuprieit]	a.	适当的，恰当的，合适的
attainment [əˈteinmənt]	n.	获得，达到；技能，才能；造诣，成就
bachelor [ˈbætʃələ]	n.	学士，学士学位；单身汉
calculus [ˈkælkjuləs]	n.	微积分，微积分学；计算
civic [ˈsivik]	a.	市民的，公民的，平民的，民事的；城市的
consultant [kənˈsʌltənt]	n.	顾问，咨询者，请教者，商议者
commerce [ˈkɔməːs]	n.	商业，贸易
component [kəmˈpəunənt]	a.	组成的，合成的，构成的
	n.	组成部分，成分，要素
comprehensive [ˌkɔmpriˈhensiv]	a.	广泛的，综合的，全面的
comprise [kəmˈpraiz]	v.	包括，包含，由……组成
computational [kəmpjuːˈteiʃənəl]	a.	计算的，计算机的
conjunction [kənˈdʒʌŋkʃən]	n.	连接，联结，结合，联合
constraint [kənˈstreint]	n.	监禁，关押；限制，抑制，压制
contemporary [kənˈtempərəri]	a.	同时代存在的，属于同一时期的，同年龄的；现代的，当代的
context [ˈkɔntekst]	n.	上下文，语境，（事件、人物等的）来龙去脉，背景
criteria [kraiˈtiəriə]	n.	（批评、判断的）标准，准则，尺度（单数形式为 criterion）

critical	['kritikəl]	a.	爱挑剔的，吹毛求疵的；批评的，批判的，评论性的；紧要的，决定性的，关键的；危险的，危及的，危机的
cross-cutting	['krɔsˌkʌtiŋ]	n.	横切，交叉剪接
		a.	横切的，交叉的
culminate	['kʌlmineit]	v.	使告终，使结束，达到顶点，达到极点（常与 in 连用）；告终（常与 in 连用）
ethics	['eθiks]	n.	伦理学，道德学，伦理观，道德体系，道德准则
ethical	['eθikəl]	a.	伦理学的，伦理上的，道德上的，（合乎）道德（标准）的
expertise	[ˌekspəː'tiːz]	n.	专门技能，专长；鉴定书
expository	[ik'spɔzitəri]	a.	阐述的，解释的，说明（性）的，讲解的
exposure	[ik'spəuʒə]	n.	显露，暴露；公开，透露
faculty	['fækəlti]	n.	能力，才能；机能，官能，功能；（大学的）系，科，学院；全体教员
formulate	['fɔːmjuleit]	v.	使公式化，用公式表示；系统地阐述，构想出（计划、方法等），规划（制度等）
hazard	['hæzəd]	n.	危险，冒险；危害物
incorporate	[in'kɔːpəreit]	v.	包含，吸收，加入；合并，使并入
in-depth	[in'depθ]	a.	深入的，十分彻底的，全面的，详细的
interpret	[in'təːprit]	v.	解释，说明，阐明；理解，了解
licensure	['laisənʃuə]	n.	许可证（如营业执照等）的颁发
macroeconomic	[ˌmækrəuˌiːkə'nɔmik]	a.	宏观经济学的
magnetics	[mæg'netiks]	n.	磁学
microeconomic	[ˌmaikrəuˌiːkə'nɔmik]	a.	微观经济学的
minority	[mai'nɔrəti]	n.	少数，少数民族，少数派

Unit 4 Civil Engineering Major

multidisciplinary	[ˌmʌlti'disiplinəri]	a.	包括（涉及）多种学科的
multiple	['mʌltipl]	a.	多个的，由多个组成的，多重的，多种多样的，复杂的
orientation	[ˌɔːrien'teiʃn]	n.	定位，定向；方位，方向
professionalism	[prəu'feʃənəlizəm]	n.	专业性，职业特性，职业作风，职业精神
prominent	['prɔminənt]	a.	突起的，凸出的；突出的，显著的，杰出的，重要的
resilient	[ri'ziliənt]	a.	有回弹力的，有弹性的，能迅速恢复原状的
submission	[səb'miʃn]	n.	屈服，服从，归顺，投降；提交物，呈递物
substantial	[səb'stænʃəl]	a.	大量的，实质的
withstand	[wið'stænd]	v.	反对，对抗，反抗；经得起，顶得住

Proper Names

BTW (business and technical writing)	商务和技术写作
CEE (civil and environmental engineering)	土木与环境工程
CHEM (Chemistry)	化学
CS (computing science)	计算科学
differential equation	微分方程
ECON (economics)	经济学
ENG (engineering)	工程学
GE (graphics engineering)	图形工程
graduate school	研究生院
MATH (mathematics)	数学
matrix theory	矩阵理论

PHYS (physics)　　　　　　　　　　　　　　物理学
RHET (rhetoric studies)　　　　　　　　　　修辞学研究
standing committee　　　　　　　　　　　　常务委员会
TAM (theoretical and applied mechanics)　　理论与应用力学
University of Illinois at Urbana-Champaign　伊利诺伊大学香槟分校

I True and false.

1. The professional component of each engineering program includes one year of a combination of college-level mathematics and basic sciences, one and one-half years of engineering topics, and a general education component. (□T　□F)
2. Civil engineering works usually require cooperation among professionals of many different disciplines. (□T　□F)
3. Each area of the engineering program is based on a different fundamental core principle. (□T　□F)
4. The courses of orientation and professional development stress fundamental concepts and provide the skills to work effectively and successfully in the engineering profession. (□T　□F)
5. Writing and Research is a course designed to teach fundamentals of argumentative writing. (□T　□F)

II Choose the best answer according to the text.

1. Which of the following statements about engineering programs is incorrect?
   A. The educational objectives of each engineering program are based on the mission and vision statement of the college.
   B. The educational objectives describe what graduates are expected to attain within a few years of graduation.
   C. From the programs students receive a comprehensive education that includes in-depth instruction in their chosen fields of study.

D. The programs are designed to emphasize technical knowledge and skills and to provide exposure to open-ended problems and design methods.

2. Which of the following program is not included in the cross-cutting programs?
   A. Societal risk management.
   B. Energy, water, and environmental sustainability.
   C. Transportation engineering.
   D. Sustainable and resilient infrastructure systems.

3. Which of the following statement about the purpose of the review and approval process of each student's academic program plan is incorrect?
   A. To ensure that individual programs satisfy the educational objectives and all of the requirements of the civil engineering program.
   B. To ensure that students can continue their formal education at a graduate school of their choice.
   C. To ensure that those programs do not abuse the substantial degree of flexibility present in the curriculum.
   D. To ensure that the career interests of each student are cultivated and served.

4. Which of the following subjects is not offered to stress the basic mathematical and scientific principles?
   A. Chemistry.   B. Biology.   C. Mathematics.   D. Physics.

5. Which of the following courses is not included in the civil engineering technical electives?
   A. Engineering Risk and Uncertainty.
   B. Construction Materials Engineering.
   C. Geotechnical Engineering.
   D. Societal Risk Management.

III Translation.

1. 这些工程专业使毕业生能够在他们所选择的领域做出重大贡献，并同时意识到他们的社会责任。
   A. The engineering programs enable graduates make significant contributions in their choice fields while at the same time recognizing their responsibilities to society.
   B. The engineering programs enable graduates to make significant contributions in

their choice fields while at the same time fulfilling their responsibilities to society.

C. The engineering programs enable graduates to make significant contributions in their chosen fields while at the same time recognizing their responsibilities to society.

D. The engineering programs enable graduates make significant contributions in their chosen fields while at the same time fulfilling their responsibilities to society.

2. 这门选修课能够提高学生科学知识的深度。

A. This elective allows the student to gain addition depth in science.

B. This elective allows the student to gain additional depth in science.

C. This elective allows the student to gain additional deep in science.

D. This elective allows the student to gain deep addition in science.

3. 文科教育课程增进学生对人类文化和社会的了解，培养学生的探究精神和批判性思维，并为学生的公民参与行为和终身学习奠定基础。

A. The liberal education courses develop students' understanding of human culture and society, build skills of inquiry and critical thinking, and lay a foundation for civic engagement and lifelong learning.

B. The liberal education courses develop students' understanding of human culture and society, build skills of inquiry and critical thought, and lay a foundation for civic engagement and lifetime learning.

C. The liberal education courses develop students' understanding of human culture and society, build skills of inquiry and critical thinking, and lay a foundation for civilized engagement and lifetime learning.

D. The liberal education courses develop students' understanding of human culture and society, build skills of inquiry and critical thought, and lay a foundation for civilized engagement and lifelong learning.

4. 大多数学生选择能够同时满足这些文化学习要求的文科教育课程。

A. Most students elect liberal education courses that spontaneously satisfy these cultural studies requirements.

B. Most students select liberal education courses that simultaneously satisfy these cultural studies requirements.

C. Most students select liberal education courses that spontaneously satisfy these cultural studies demands.

D. Most students elect liberal education courses that simultaneously satisfy these cultural studies demands.

Section C Supplementary Reading

The Department of Civil Engineering of the University of Hong Kong

Since the establishment of the University of Hong Kong and the Faculty of Engineering in 1912, the Department of Civil Engineering has nurtured many brilliant leaders in the civil engineering discipline and made significant contributions to the local and overseas community. The Department is constantly looking ahead to enhancing its goals in education, research and community services in order to keep abreast of the ever-changing demands of modern society. We are very pleased to have ranked 9th globally for the second year under the QS University Subject Rankings 2016 in the subject area of civil engineering.

The Department has switched to a new and innovative 4-year curriculum in 2012, to equip students with knowledge beyond traditional civil engineering coupled with versatile options, for example the double degree and the major/minor. In our new 4-year curriculum, greater emphasis is placed on experiential learning in the form of project-based design, where students participate in engineering projects relevant to their fields of study. A new curriculum development plan has also been developed in the new triennium and from the 2017–2018 academic year, we would introduce new "specializations" for students, to allow them to focus on certain areas for more in-depth studies.

Since 2004, the Department has also established the Project Mingde, where arrangements are made for our students to take up the design and construction of real life projects in China. The first project—Mingde Building, a primary school, was built in Guangxi Province in 2005 and the second project — Gewu Building, a dormitory for the Rongshui Vocational Training School, Guangxi Province, was completed in 2008. The third project, a kindergarten in Chongzhou,

Sichuan Province was completed in November 2011. The fourth project was building of Chaoyang Bridge at Yingdong Village, Guangxi Province and it was completed in June 2013. The fifth project, a Community and Cultural Centre at Dabao Village, Guangxi Province was completed in May 2015. In 2016, the latest project of Project Mingde is to build a teacher accommodation for Daping Primary School in Guangxi Province. We would continue to look for meaningful and educational projects and opportunities for our students.

Through the Mingde Projects, the educational goal of bringing real projects into the classroom and, vice versa, bringing the classroom into the projects, is realized. The Department is very fortunate to have a group of dedicated alumni to provide professional guidance and mentorship to our students for these various projects.

The Department has continuously attracted top students and earned a good reputation in both academic and industry. With the government's implementation of mega-infrastructure projects, there is a great demand for civil engineers. Moreover, the construction boom in the Mainland, the Middle East and beyond, also opened up new opportunities for young and enthusiastic civil engineers to participate in the boundless infrastructure developments in China and overseas. The Department of Civil Engineering will continue to devote itself to teaching and research for the betterment of society.

The Departmental Mission Statement

To produce highly qualified, well-rounded, and motivated graduates possessing fundamental knowledge of civil engineering who can provide leadership and professional service to Hong Kong, Mainland China and the World.

To pursue creative research and new technologies in civil engineering and across disciplines in order to serve the needs of the industry, government, society, and the scientific community by expanding the knowledge in the field.

To develop partnerships with industrial and government agencies.

Unit 4　Civil Engineering Major

　　To achieve visibility by active participation in international and local conferences as well as relevant technical and community activities.

Civil Engineering in Hong Kong

　　Hong Kong has enjoyed tremendous growth in the past three decades. Such enviable growth and economic achievement have been made possible through the success of infrastructure projects—the cross harbour tunnels, the mass transit railways, the new towns, the highways, the water supply and wastewater treatment systems.  The new airport at Chek Lap Kok, the new container port at Tsing Yi and Lantau, the new Route 3 and the western corridor railway to the border, as well as massive sewage collection, treatment and disposal works are now under construction. These projects will ensure future growth as well as improvement in our quality of life. Engineering graduates from the University of Hong Kong, particularly the civil engineering graduates, have made and will continue to make the most significant contributions in all of these developments.

The World University Rankings in Civil Engineering Major (2014)

Ranking	Name	Nation/Region
1	Massachusetts Institute of Technology (MIT)	USA
2	University of Illinois at Urbana-Champaign	USA
3	University of California, Berkeley (UCB)	USA
4	The University of Tokyo	Japan
4	University of Cambridge	UK
6	University of Texas at Austin	USA
7	National University of Singapore (NUS)	Singapore
8	Kyoto University	Japan
9	Imperial College London	UK
10	University of Hong Kong	HK, China
11	Stanford University	USA
12	The Hong Kong University of Science and Technology	HK, China
13	Swiss Federal Institute of Technology Zurich (ETH Zurich)	Switzerland
14	Delft University of Technology	Netherlands

续表

Ranking	Name	Nation/Region
15	The Hong Kong Polytechnic University	HK, China
15	The University of Sydney	Australia
17	Politecnico di Milano	Italy
18	The University of New South Wales	Australia
19	University of Canterbury	New Zealand
20	Tsinghua University	China
21	Nanyang Technological University (NTU)	Singapore
21	Tongji University	China
23	Purdue University	USA
24	Ecole Polytechnique Fédérale de Lausanne	Switzerland
24	Georgia Institute of Technology	USA
26	Texas A&M University	USA
27	University of California, San Diego (UCSD)	USA
28	National Technical University of Athens	Greece
29	Tokyo Institute of Technology	Japan
30	Monash University	Australia
31	City University of Hong Kong	HK, China
32	KAIST-Korea Advanced Institute of Science & Technology	Korea
33	University of Toronto	Canada
34	University of California, Davis	USA
35	Taiwan University	Taiwan, China
36	Universitat Politècnica de Catalunya	Spain
37	Cornell University	USA
38	Seoul National University	Korea
39	University of Oxford	UK
40	Karlsruhe Institute of Technology (KIT)	Germany
41	Politecnico di Torino	Italy
42	California Institute of Technology (Caltech)	USA
42	The University of Queensland	Australia
44	The University of Melbourne	Australia
45	The University of Newcastle	Australia

续表

Ranking	Name	Nation/Region
46	KTH, Royal Institute of Technology	Sweden
47	Queensland University of Technology	Australia
48	Technical University of Denmark	Denmark
49	Indian Institute of Technology Madras (IITM)	India
50	Indian Institute of Technology Bombay (IITB)	India
50	University of British Columbia	Canada

Unit 5
Building Materials

Section A Text

Building Materials

Building materials are any materials which are used for construction purposes. Many naturally occurring substances, such as clay, rocks, sand, and wood, even twigs and leaves, have been used to construct buildings. Apart from naturally occurring materials, many man-made products are in use, some more and some less synthetic. The manufacture of building materials is an established industry in many countries and the use of these materials is typically segmented into specific specialty trades, such as carpentry, insulation, plumbing, and roofing work. They provide the make-up of habitats and structures including homes.

In history there are trends in building materials from being natural to becoming more man-made and composite; biodegradable to imperishable; indigenous to being transported globally; repairable to disposable; chosen for increased levels of fire-safety and improved seismic resistance.

Unit 5 Building Materials

Naturally Occurring Substances

1. Mud and clay

Clay based buildings usually come in two distinct types: the walls of one type are made directly with the mud mixture, and the walls of the other type are built by stacking air-dried building blocks called mud bricks. Other use of clay in building is combined with straws to create light clay, wattle and daub, and mud plaster. People building with mostly dirt and clay, such as cob, sod, and adobe, created  homes that have been built for centuries in western and northern Europe, Asia, as well as the rest of the world, and continue to be built, though on a smaller scale. Some of these buildings have remained habitable for hundreds of years.

2. Sand

Sand is used with cement, and sometimes lime, to make mortar for masonry work and plaster. Sand is also used as a part of the concrete mix.

3. Stone or rock

Rock structures have existed for as long as history can recall. It is the longest lasting building material available, and is usually readily available. There are many types of rock throughout the world, all with differing attributes that make them better or worse for particular uses. Rock is a very dense material so it gives a lot of protection too; its main drawback as a material is its weight and awkwardness. Its energy density is also considered a big drawback, as stone is hard to keep warm without using large amounts of heating  resources. Dry-stone walls have been built for as long as humans have put one stone on top of another. Eventually, different forms of mortar are used to hold the stones together, cement being the most commonplace now. Stone buildings can be seen in most major cities; some civilizations built entirely with stone such as the Egyptian and Aztec pyramids and the structures of the Inca civilization.

4. Thatch

Thatch is one of the oldest of building materials known; grass is a good insulator and easily harvested. Many African tribes have lived in homes made completely of grasses and sand year-round. In Europe, thatch roofs on homes were once prevalent but the material fell out of favor as industrialization and improved transport increased the availability of other materials. Today, though, the practice is undergoing a revival. In the Netherlands, for instance, many new buildings have thatched roofs with special ridge tiles on top.

5. Wood and timber

Wood has been used as a building material for thousands of years in its natural state. Today, engineered wood is becoming very common in industrialized countries. Wood is a product of trees, and sometimes other fibrous plants, used for construction purposes when cut or pressed into lumber and timber, such as boards, planks and similar materials. It is a generic building material and is used in building just about any type of structure in most climates. Wood can be very flexible under loads, keeping strength while bending, and is incredibly strong when compressed vertically. There are many differing qualities to the different types of wood, even among the same tree species. This means specific species are better suited for various uses than others. And growing conditions are important for deciding quality. "Timber" is the term used for construction purposes except the term "lumber" is used in the United States. Raw wood (a log, trunk, bole) becomes timber when the wood has been "converted" (sawn, hewn, split) in the forms of minimally-processed logs stacked on top of each other, timber frame construction, and light-frame construction. The main problems with timber structures are fire risk and moisture-related problems.

Man-made Substances

1. Fired bricks and clay blocks

Bricks are made in a similar way to mud-bricks except without the fibrous binder such as straw and are fired after they have air-dried to permanently harden them. Fired bricks can be solid or have hollow cavities to aid in drying and make them lighter and easier to transport. The individual bricks are placed upon each other in courses using mortar. Successive courses being used to build up walls, arches, and other architectural elements. Fired brick walls are usually substantially thinner than adobe while keeping the same vertical strength. They require more energy to create but are easier to transport and store, and are lighter than stone blocks. Romans extensively used fired brick of a shape and type now called Roman bricks. Building with bricks gained much popularity in the mid-18th century and 19th centuries. This  was due to lower costs with increases in brick manufacturing and fire-safety in the ever crowding cities. The cinder block supplemented or replaced fired bricks in the late 20th century often being used for the inner parts of masonry walls and by themselves. Structural clay tiles (clay blocks) are clay or terracotta and typically are perforated with holes.

2. Cement composites

Cement bonded composites are made of hydrated cement paste that binds wood, particles, or fibers to make pre-cast building components. Various fibrous materials, including paper, fiberglass, and carbon-fiber have been used as binders. Wood and natural fibers are composed of various soluble organic compounds like carbohydrates, glycosides and phenolic. These compounds are known to retard cement setting. Therefore, before using a wood in making cement bonded composites, its compatibility with cement is assessed. Wood-cement compatibility is the ratio of a parameter related to the property of a wood-cement composite to that of a neat cement paste. The compatibility is often expressed as a percentage value. To determine wood-cement compatibility, methods based on different properties are used, such as hydration characteristics, strength, interfacial bond and morphology. Various methods are used by researchers such as the measurement of hydration characteristics of a cement-aggregate mix; the comparison of the mechanical properties of cement-

aggregate mixes and the visual assessment of micro-structural properties of the wood-cement mixes. It has been found that the hydration test by measuring the change in hydration temperature with time is the most convenient method.

3. Concrete

Concrete is a composite building material made from the combination of aggregate and a binder such as cement. The most common form of concrete is Portland cement concrete, which consists of mineral aggregate (generally gravel and sand), Portland cement and water. After mixing, the cement hydrates and eventually hardens into a stone-like material. When used in the generic sense, this is the material referred to by the term "concrete". For a concrete construction of any size, as concrete has a rather low tensile strength, it is generally strengthened using steel rods or bars. This strengthened concrete is then referred to as reinforced concrete. In order to minimize any air bubbles, which would weaken the structure, a vibrator is used to eliminate any air that has been entrained when the liquid concrete mix is poured around the ironwork. Concrete has been the predominant building material in the modern age due to its longevity, formability, and ease of transport.

4. Fabric

The tent is the home of choice among nomadic groups all over the world. Two well-known types include the conical teepee and the circular yurt. The tent has been revived as a major construction technique with the development of tensile architecture and synthetic fabrics. Modern buildings can be made of flexible material such as fabric membranes, and supported by a system of steel cables, rigid or internal, or by air pressure.

5. Foam

Recently, synthetic polystyrene foam has been used in combination with structural materials, such as concrete. It is lightweight, easily shaped, and an excellent insulator. Foam is usually used as part of a structural insulated panel, wherein the foam is sandwiched between wood, cement or insulating concrete forms.

6. Glass

Glassmaking is considered an art form as well as an industrial process or material. Clear windows have been used since the invention of glass to cover small openings in a building. Glass panes provide humans with the ability to both let light into rooms while at the same time keeping inclement weather outside. Glass is generally made from mixtures of sand and silicates, in a very hot fire stove called a kiln, and it is very brittle. Additives are often included the mixture used to produce glass with shades of colors or various characteristics, such as bulletproof glass or light bulbs.  The use of glass in architectural buildings has become very popular in the modern culture. Glass "curtain walls" can be used to cover the entire facade of a building, or it can be used to span over a wide roof structure in a "space frame". These uses though require some sort of frame to hold sections of glass together, as glass by itself is too brittle and would require an overly large kiln to be used to span such large areas. Glass bricks were invented in the early 20th century.

7. Metal

Metal is used as structural framework for larger buildings such as skyscrapers, or as an external surface covering. There are many types of metals used for building. Metal figures quite prominently in prefabricated structures such as the Quonset hut, and can be seen used in most cosmopolitan cities. It requires a great deal of human labor to produce metal, especially in the large amounts needed for the building industries. Corrosion is metal's prime enemy  when it comes to longevity. Steel is a metal alloy whose major component is iron, and is the usual choice for metal structural building materials. It is strong, flexible, and if refined well and/or treated lasts a long time. The lower density and better corrosion resistance of aluminum alloys and tin sometimes overcome their greater cost. Copper is a valued building material because of its advantageous properties. These include corrosion resistance, durability, low thermal movement, light weight, radio frequency shielding, lightning protection, sustainability, recyclability, and a wide range of finishes. Copper is incorporated into roofing, flashing, gutters, downspouts, domes, spires, vaults, wall cladding, building expansion joints, and indoor design elements. Other metals used include chrome, gold, silver, and titanium. Titanium can be used for structural purposes, but it is much more expensive than steel. Chrome, gold, and silver are used as decoration, because these materials are expensive and lack structural qualities such as tensile strength or hardness.

8. Plastics

The term "plastics" covers a range of synthetic or semi-synthetic organic condensation or polymerization products that can be molded or extruded into objects, films, or fibers. Their name is derived from the fact that in their semi-liquid state they are malleable, or have the property of plasticity. Plastics vary immensely in heat tolerance, hardness, and resiliency. Combined with this adaptability, the general uniformity of composition and lightness of plastics ensures their use in almost all industrial applications today. High performance plastics have become an ideal building material due to its high abrasion resistance and chemical inertness. Notable buildings that feature it include: the Beijing National Aquatics Center and the Eden Project biomes.

9. Papers and membranes

Building papers and membranes are used for many reasons in construction. Tar paper was invented late in the 19th century and was used as rosin paper and for gravel roofs. Tar paper has largely fallen out of use supplanted by asphalt felt paper. Felt paper has been supplanted in some uses by synthetic underlayment, particularly in roofing by synthetic underlayment and siding by house wraps. There are a wide variety of damp proofing and waterproofing membranes used for roofing, basement waterproofing, and geomembranes.

Unit 5 Building Materials

Section B Text Exploration

New Words and Expressions

adobe [ə'dəubi]	n.	土砖，砖坯
alloy ['æloi; ə'loi]	n.	合金
biodegradable [ˌbaiəudi'greidəbl]	a.	生物所能分解的，能进行生物降解的
biome ['baiəum]	n.	生物群系，生物群落区
brittle ['britl]	a.	易碎的，脆弱的；易生气的
carbohydrate [ˌkɑːbəu'haidreit]	n.	碳水化合物，糖类
cavity ['kævəti]	n.	腔；洞，凹处
cement [si'ment]	n.	水泥，接合剂
chrome [krəum]	n.	铬，铬合金，铬黄
cob [kɔb]	n.	玉米棒子芯，玉米穗轴
	v.	捣碎
compatibility [kəmˌpætə'biləti]	n.	兼容性
cosmopolitan [ˌkɔzmə'pɔlitən]	a.	世界性的，世界主义的，四海一家的
daub [dɔːb]	v.	涂抹，乱画，弄脏
	n.	涂抹；涂料；拙劣的画
density ['densəti]	n.	密度
downspout ['daunspaut]	n.	落水管（将雨水从屋顶排至水沟）
extrude [ek'struːd]	v.	挤出，压出，使突出，逐出，突出，喷出
	v.	用碎石铺；使船搁浅在沙滩上；使困惑
glycoside ['glaikəusaid]	n.	糖苷，苷，糖苷类
gravel ['grævəl]	n.	碎石，砂砾

gutter	['gʌtə]	n.	排水沟，槽；贫民区
hydrated	['haidreitid]	a.	含水的
hydration	[hai'dreiʃən]	n.	水合作用
imperishable	[im'periʃəbl]	a.	不朽的，不灭的
indigenous	[in'didʒinəs]	a.	本土的，土著的，国产的；固有的
insulator	['insjuleitə; 'insəleitə]	n.	绝缘体；从事绝缘工作的工人
insulation	[ˌinsju'leiʃən]	n.	隔离，绝缘，隔热，隔音；绝缘（或隔热、隔音）材料
interfacial	[ˌintə'feiʃəl]	a.	界面的
kiln	[kiln; kil]	n.	（砖、石灰等的）窑，炉，干燥炉
		v.	烧窑，在干燥炉干燥
lime	[laim]	n.	石灰；酸橙；绿黄色
lumber	['lʌmbə]	n.	木材
		v.	砍伐木材；乱堆
malleable	['mæliəbl]	a.	（金属）可锻的，可塑的，有延展性的，有韧性的；易适应的；顺从的，温顺的，可训的
membrane	['membrein]	n.	膜，薄膜，羊皮纸
morphology	[mɔː'fɔlədʒi]	n.	形态学，形态论；词法，词态学
parameter	[pə'ræmitə]	n.	参数，系数，参量
perforate	['pəːfəreit]	v.	穿孔于，打孔穿透，在……上打齿孔，穿过，贯穿，穿孔
phenolic	[fi'nɔlik]	n.	酚醛树脂
		a.	酚的，酚醛树脂的
polystyrene	[ˌpɔli'staiəriːn; ˌpɔli'staiərin]	n.	聚苯乙烯
prefabricate	[ˌpriː'fæbrikeit]	v.	预先制造，预先构思；预制构件

Unit 5 Building Materials

retard [ri'tɑːd]	v.	延迟，阻止，妨碍，使减速，减慢，受到阻滞
	n.	延迟，阻止
segment ['segmənt; seg'ment]	v.	分割
silicate ['silikit; 'silikeit]	n.	硅酸盐
sod [sɔd]	n.	草地，草皮
soluble ['sɔljubl]	a.	可溶的，可溶解的；可解决的
synthetic [sin'θetik]	a.	综合的，合成的，人造的
	n.	合成物
timber ['timbə]	n.	木材，木料
titanium [tai'teiniəm; ti'teiniəm]	n.	钛（一种金属元素）
vertically ['vəːtikəli]	ad.	垂直地
vibrator [vai'breitə]	n.	振动器，振子，振动按摩器，振动滚筒；簧片
wattle ['wɔtl]	n.	板条，编条

abrasion resistance	耐磨性，抗磨损性
asphalt felt paper	沥青油毡纸
cement composite	水泥复合材料
cement hydrate	水泥水化产物
cinder block	煤渣砖，煤渣砌块
circular yurt	圆形圆顶帐篷
concrete mix	混凝土混合物
conical teepee	圆锥形帐篷
corrosion resistance	耐蚀性，抗腐蚀性
mud plaster	泥浆石膏
plumbing work	水管工程
radio frequency shielding	射频屏蔽
reinforced concrete	钢筋混凝土
ridge tile	脊瓦，屋脊瓦

rosin paper	松香纸
seismic resistance	耐震性，抗震稳定性
synthetic fabrics	合成纤维织物，化纤布
tar paper	沥青纸，焦油沥青毡，焦油纸，防潮纸
tensile strength	抗张强度，抗拉强度，拉伸强度
wattle and daub	抹灰篱笆墙

Exercises

I True and false.

1. Rock structures have existed for a long history. (□T □F)
2. Many African tribes have lived in homes made completely of stone or rock year-round. (□T □F)
3. Fired bricks can be solid or have hollow cavities to aid in drying and make them lighter and easier to transport. (□T □F)
4. After mixing, the cement hydrates and eventually hardens into a stone-like material. (□T □F)
5. High performance plastics have become an ideal building material due to its high abrasion resistance and chemical inertness. (□T □F)

II Choose the best answer according to the text.

1. What are the trends in building materials in history?
   A. They are from being natural to becoming more man-made and composite.
   B. They are from being imperishable to becoming biodegradable.
   C. They are from being transported globally to becoming indigenous.
   D. They are from being disposable to becoming repairable.

2. _____ is the longest lasting building material available.
   A. Mud   B. Sand   C. Stone   D. Wood
3. What are the main problems of timber structures?
   A. They are expensive and lack of structural qualities.
   B. They are fragile and easily to be broken by external forces.
   C. They can never change their forms and are easily broken.
   D. They usually have fire risk and moisture-related problems.
4. Building with bricks gained much popularity in _____.
   A. the mid-17th century and 18th centuries
   B. the mid-18th century and 19th centuries
   C. the mid-19th century and 20th centuries
   D. the mid-20th century and 21th centuries
5. Clear windows have been used since the invention of _____.
   A. fabric   B. foam   C. glass   D. metal

III Translation.

1. 在欧洲，房子上的茅草屋顶曾经非常流行，然而随着工业化的进展和交通的改善，其他材料的可用性得到提高，茅草屋顶失宠了。然而，如今这种屋顶正在复兴，例如在荷兰，许多新建筑都有茅草屋顶，上面有特殊的脊瓦。

   A. In Europe, thatch roofs on homes were once prevalent but the material was out of date as industrialization and improved transport increased the stability of other materials. Until today the practice is also popular. In the Netherlands, for instance, many new buildings have thatched roofs with special ridge tiles on top.

   B. In Europe, thatch roofs on homes were once prevalent but the material fell out of favor as industrialization and improved transport increased the availability of other materials. Today, though, the practice is undergoing a revival. In the Netherlands, for instance, many new buildings have thatched roofs with special ridge tiles on top.

   C. In Europe, thatch roofs on homes were once prevalent but the material was out of date as industrialization and improved transport increased the availability of other materials. While today the practice is undergoing a crisis. In the Netherlands, for example, many new buildings refuse to have thatched roofs with special ridge tiles on top.

D. In Europe, thatch roofs on homes were once popular but the material fell out of favor as industrialization and improved transport increased the productivity of other materials. But today thatch roofs are popular again. In the Netherlands, for instance, many new buildings have thatched roofs with special ridge tiles on top.

2. 当木头"转变"成为堆叠在一起的轻度加工的木料,或者木材框架结构及轻型框架结构等形式,原木就成了木材。

A. When the wood has been "changed" into the forms of stacking logs with maximum processing, or timber frame construction and light-frame construction, raw wood becomes timber.

B. When the wood has been "changed" into the forms of stacking logs with maximum processing, or timber frame construction and light-frame construction, timber becomes lumber.

C. Timber becomes lumber only when the wood has been "converted" in the forms of minimally-processed logs stacked on top of each other, timber frame construction, and light-frame construction.

D. Raw wood becomes timber when the wood has been "converted" in the forms of minimally-processed logs stacked on top of each other, timber frame construction, and light-frame construction.

3. 混凝土是由骨料和诸如水泥之类的黏合剂组合而成的复合建筑材料。最常见的混凝土是波特兰水泥混凝土,它由矿物骨料、波特兰水泥和水组成。

A. Concrete is made from the combination of bone and a binder such as cement, and it is a composite building material. The most common form of concrete is Portland cement concrete, which consists of mineral bone, Portland cement and water.

B. Cement is made from the combination of bone and a binder such as concrete, and it is a composite building material. The most common form of cement is Portland cement, which consists of mineral bone, Portland cement and water.

C. Concrete is a composite building material made from the combination of aggregate and a binder such as cement. The most common form of concrete is Portland cement concrete, which consists of mineral aggregate, Portland cement and water.

D. Cement is a composite building material made from the combination of aggregate and a binder such as concrete. The most common form of cement is Portland cement, which consists of mineral aggregate, Portland cement and water.

Section C Supplementary Reading

Wood

Wood is a porous and fibrous structural tissue found in the stems and roots of trees and other woody plants. It is an organic material, a natural composite of cellulose fibers that are strong in tension and embedded in a matrix of lignin that resists compression. Wood is sometimes defined as only the secondary xylem in the stems of trees, or it is defined more broadly to include the same type of tissue elsewhere such as in the roots of trees or shrubs. In a living tree it performs a support function, enabling woody plants to grow large or to stand up by themselves. It also conveys water and nutrients between the leaves, other growing tissues, and the roots. Wood may also refer to other plant materials with comparable properties, and to material engineered from wood, or wood chips or fiber.

Wood has been used for thousands of years for fuel, as a construction material, for making tools and weapons, furniture and paper, and as a feedstock for the production of purified cellulose and its derivatives, such as cellophane and cellulose acetate.

In 2005, the growing stock of forests worldwide was about 434 billion cubic meters, 47% of which was commercial. As an abundant, carbon-neutral renewable resource, woody materials have been of intense interest as a source of renewable energy. In 1991 approximately 3.5 billion cubic meters of wood were harvested. Dominant uses were for furniture and building construction.

Wood Density

Wood density is determined by multiple growth and physiological factors compounded into "one fairly easily measured wood characteristic". Age, diameter, height, radial (trunk) growth, geographical location, site and growing conditions, silvicultural treatment, and seed source all to some degree influence wood density. Variation is to be expected. Within an individual tree, the variation in wood density is often as great as or even greater than that between different trees. Variation of specific gravity within the bole of a tree can occur in either the horizontal or vertical direction.

Hard and Soft Woods

It is common to classify wood as either softwood or hardwood. The wood from conifers (e. g. pine) is called softwood, and the wood from dicotyledons (usually broad-leaved trees, e. g. oak) is called hardwood. These names are a bit misleading, as hardwoods are not necessarily hard, and softwoods are not necessarily soft. The well-known balsa (a hardwood) is actually softer than any commercial softwood. Conversely, some softwoods (e. g. yew) are harder than many hardwoods.

There is a strong relationship between the properties of wood and the properties of the particular tree that yielded it. The density of wood varies with species. The density of a wood correlates with its strength (mechanical properties). For example, mahogany is a medium-dense hardwood that is excellent for fine furniture crafting, whereas balsa is light, making it useful for model building. One of the densest woods is black ironwood.

Wood Flooring

Wood can be cut into straight planks and made into a wood flooring. A solid wood floor is floor laid with planks or battens which have been created from a single piece of timber, usually a hardwood. Since wood is hydroscopic (it acquires and loses moisture from the ambient conditions around it) this potential instability effectively limits the length and width of the boards.

Solid hardwood flooring is usually cheaper than engineered timbers and damaged areas can be sanded down and refinished repeatedly, the number of times being limited only by the thickness of wood above the tongue.

Solid hardwood floors were originally used for structural purposes, being installed perpendicular to the wooden support beams of a building (the joists or bearers) and solid construction timber is still often used for sports floors as well as most traditional wood blocks, mosaics and parquetry.

Engineered Wood

Engineered wood products, glued building products "engineered" for application-specific performance requirements, are often used in construction and industrial applications. Glued engineered wood products are manufactured by bonding together wood strands, veneers, lumber or other forms of wood fiber with glue to form a larger, more efficient composite structural unit.

These products include glued laminated timber (glulam), wood structural panels (including plywood, oriented strand board and composite panels), laminated veneer lumber (LVL) and other structural composite lumber (SCL) products, parallel strand lumber, and I-joists. Approximately 100 million cubic meters of wood was consumed for this purpose in 1991. The trends suggest that particle board and fiber board will overtake plywood.

Wood unsuitable for construction in its native form may be broken down mechanically (into fibers or chips) or chemically (into cellulose) and used as a raw material for other building materials, such as engineered wood, as well as chipboard, hardboard, and medium-density fiberboard (MDF). Such wood derivatives are widely used: wood fibers are an important component of most paper, and cellulose is used as a component of some synthetic materials. Wood derivatives can also be used for kinds of flooring, for example laminate flooring.

Furniture and Utensils

Wood has always been used extensively for furniture, such as chairs and beds. It is also used for tool handles and cutlery, such as chopsticks, toothpicks, and other utensils, like the wooden spoon and pencil.

Construction Use

Wood has been an important construction material since humans began building shelters, houses and boats. Nearly all boats were made out of wood until the late 19th century, and wood remains in common use today in boat construction. Elm in particular was used for this purpose as it resisted decay as long as it was kept wet (it also served for water pipe before the advent of more modern plumbing).

Wood to be used for construction work is commonly known as lumber in North America. Elsewhere, lumber usually refers to felled trees, and the word for sawn planks ready for use is timber. In Medieval Europe oak was the wood of choice for all wood construction, including beams, walls, doors, and floors. Today a wider variety of woods is used: solid wood doors are often made from poplar, small-knotted pine, and Douglas fir.

New domestic housing in many parts of the world today is commonly made from timber-framed construction. Engineered wood products are becoming a bigger part of the construction industry. They may be used in both residential and commercial buildings as structural and aesthetic materials.

In buildings made of other materials, wood will still be found as a supporting material, especially in roof construction, in interior doors and their frames, and as exterior cladding. Wood is also commonly used as shuttering material to form the mold into which concrete is poured during reinforced concrete construction.

The churches of Kizhi, Russia are among a handful of World Heritage Sites built entirely of wood, without metal joints.

1. Kizhi Pogost

Kizhi Pogost is a historical site dating from the 17th century on Kizhi island. The island is located on Lake Onega in the Republic of Karelia (Medvezhyegorsky District), Russia. The pogost is the area inside a fence which includes two large wooden churches (the 22-dome Transfiguration Church and the 9-dome Intercession Church) and a bell-tower. The pogost is famous for its beauty and longevity, despite that it is built exclusively of wood. In 1990, it was included in the UNESCO (United Nations Educational, Scientific, and Cultural Organization) list of World Heritage sites and in 1993 listed as a Russian Cultural Heritage site.

2. The Church of the Transfiguration

The Church of the Transfiguration is the most remarkable part of the pogost. It is not heated and is, therefore, called a summer church and does not hold winter services. Its altar was laid June 6, 1714, as inscribed on the cross located inside the church. This church was built on the site of the old one which was burnt by lightning. The builders' names are unknown. A legend tells that the main builder used one axe for the whole construction, which he threw into the lake upon completion with the words "there was not and will be not another one to match it".

The church has 22 domes and with a height of 37 meters is one of the tallest wooden buildings of the Russian North. Its perimeter is 20 × 29 meters. It is considered that the 18-dome church on the southern shore of Lake Onega, built in 1708 and destroyed by fire in 1963, was its forerunner. According to the Russian carpentry traditions of that time, the Transfiguration Church was built of wood only with no nails. All structures were made of scribe-fitted horizontal logs, with interlocking corner joinery—either round notch or dovetail — cut by axes. The basis of the structure is the octahedral frame with four two-stage side attachments. The

eastern prirub has a pentagonal shape and contains the altar. Two smaller octagons of similar shape are mounted on top of the main octagon. The structure is covered in 22 domes of different size and shape, which run from the top to the sides. The refectory is covered with a three-slope roof. In the 19th century, the church was decorated with batten and some parts were covered with steel. It was restored to its original design in the 1950s.

The church framework rests on a stone base without a deep foundation, except for the western aisle for which a foundation was built in 1870. Most wood is pine with spruce planks on the flat roofs. The domes are covered in aspen.

The iconostasis has four levels and contains 102 icons. It is dated to the second half of the 18th-early 19th century. The icons are from three periods: the two oldest icons, "The Transfiguration" and "Pokrov" are from the late 17th century and are typical of the northern style. The central icons are from the second half of the 18th century and are also of the local style. Most icons of the three upper tiers are of the late 18th century, brought from various parts of Russia.

3. The Church of the Intercession

The Church of the Intercession is a heated church where services are held from October 1 until Easter. The church was the first on the island after a fire in the late 17th century destroyed all previous churches. It was first built in 1694 as a single-dome structure, and then reconstructed in 1720-1749 and in 1764 rebuilt into its present 9-dome design as an architectural echo of the main Transfiguration Church. It stands 32 meters tall with a 26 × 8 meter perimeter. There are nine domes, one larger in the center, surrounded by eight smaller ones. Decoration is scant. A high single-part porch leads into the four interior parts of the church. As in the Transfiguration Church, the altar is placed in the eastern part shaped as a pentagon. The original iconostasis was replaced at the end of the 19th century and is lost; it was rebuilt in the 1950s to the original style.

Unit 6
Stress and Strain

Section A Text

Stress and Strain

Stress is a physical quantity that expresses the internal forces that neighboring particles of a continuous material exert on each other, while strain is the measure of the deformation of the material. For example, when a solid vertical bar is supporting a weight, each particle in the bar pushes on the particles immediately below it. When a liquid is in a closed container under pressure, each particle gets pushed against by all the surrounding particles. The container walls and the pressure-inducing surface (such as a piston) push against them in (Newtonian) reaction. These macroscopic forces are actually the net result of a very large number of intermolecular forces and collisions between the particles in those molecules.

In some branches of engineering, the term stress is occasionally used in a looser sense as a synonym of "internal force". For example, in the analysis of trusses, it may refer to the total traction or compression force acting on a beam, rather than the force divided by the area of its cross-section.

Strain inside a material may arise by various mechanisms, such as stress as applied by external forces to the bulk material (like gravity) or to its surface (like contact forces, external pressure, or friction). Any strain (deformation) of a solid material generates an internal elastic stress, analogous to the reaction force of a spring, which tends to restore the

material to its original non-deformed state. In liquids and gases, only deformations that change the volume generate persistent elastic stress. However, if the deformation is gradually changing with time, even in fluids there will usually be some viscous stress, opposing that change. Elastic and viscous stresses are usually combined under the name mechanical stress.

Strain is a measure of deformation representing the displacement between particles in the body relative to a reference length. Deformation is the transformation of a body from a reference configuration to a current configuration. A configuration is a set containing the positions of all particles of the body. A deformation may be caused by external loads, body forces (such as gravity or electromagnetic forces), or changes in temperature, moisture content, or chemical reactions, etc.

Strain is a description of deformation in terms of relative displacement of particles in the body that excludes rigid-body motions. Different equivalent choices may be made for the expression of a strain field depending on whether it is defined with respect to the initial or the final configuration of the body and on whether the metric tensor or its dual is considered.

A general deformation of a body can be expressed in the form $x = F(X)$ where X is the reference position of material points in the body. Such a measure does not distinguish between rigid body motions (translations and rotations) and changes in shape (and size) of the body. A deformation has units of length.

A strain is in general a tensor quantity. Physical insight into strains can be gained by observing that a given strain can be decomposed into normal and shear components. The amount of stretch or compression along material line elements or fibers is the normal strain, and the amount of distortion associated with the sliding

of plane layers over each other is the shear strain, within a deforming body. This could be applied by elongation, shortening, or volume changes, or angular distortion.

The state of strain at a material point of a continuum body is defined as the totality of all the changes in length of material lines or fibers, the normal strain, which pass through that point and also the totality of all the changes in the angle between pairs of lines initially perpendicular to each other, the shear strain, radiating from this point. However, it is sufficient to know the normal and shear components of strain on a set of three mutually perpendicular directions.

If there is an increase in length of the material line, the normal strain is called tensile strain, otherwise, if there is reduction or compression in the length of the material line, it is called compressive strain.

The relation between mechanical stress, deformation, and the rate of change of deformation can be quite complicated, although a linear approximation may be adequate in practice if the quantities are small enough. Stress that exceeds certain strength limits of the material will result in permanent deformation (such as plastic flow, fracture, and cavitation) or even change its crystal structure and chemical composition.

Engineering Strain

The engineering strain is expressed as the ratio of total deformation to the initial dimension of the material body in which the forces are being applied. The true shear strain is defined as the change in the angle (in radians) between two material line elements initially perpendicular to each other in the undeformed or initial configuration. The engineering shear strain is defined as the tangent of that angle, and is equal to the length of deformation at its maximum divided by the perpendicular length in the plane of force application which sometimes makes it easier to calculate.

Stress Strain Curve

The relationship between the stress and strain that a particular material displays is known as that particular material's stress strain curve. It is unique for each material and is found by recording the amount of strain at distinct intervals of tensile or compressive loading. These curves reveal many of the properties of a material.

Stress strain curves of various materials vary widely, and different tensile tests conducted on the same material yield different results, depending upon the temperature of the specimen and the speed of the loading. It is possible, however, to distinguish some common characteristics among the stress strain curves of various groups of materials and, on this basis, to divide materials into two broad categories; namely, the ductile materials and the brittle materials.

Consider a bar of cross sectional area A being subjected to equal and opposite forces F pulling at the ends so the bar is under tension. The material is experiencing a stress defined to be the ratio of the force to the cross-sectional area of the bar: $stress = F/A$. This stress is called the tensile stress because every part of the object is subjected to tension. The SI unit of stress is the newton per square meter, which is called the pascal. 1 pascal = 1 Pa = 1 N/m². Now consider a force that is applied tangentially to an object. The ratio of the shearing force to the area A is called the shear stress. If the object is twisted through an angle q, then the shear strain is: $strain = \tan q$. Finally, the shear modulus (MS) of a material is defined as the ratio of shear stress to shear strain at any point in an object made of that material. The shear modulus is also known as the torsion modulus.

Ductile Materials

Ductile materials, which include structural steel and many alloys of other metals, are characterized by their ability to yield at normal temperatures.

Low carbon steel generally exhibits a very linear stress strain relationship up to a well-defined yield point (Fig. 6 – 1).

Unit 6　Stress and Strain

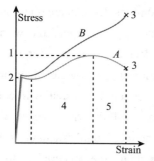

A stress strain curve typical of structural steel.
1: Ultimate strength
2: Yield strength (yield point)
3: Rupture
4: Strain hardening region
5: Necking region
A: Apparent stress ($F/A0$)
B: Actual stress (F/A)

Fig. 6−1

　　The linear portion of the curve is the elastic region and the slope is the modulus of elasticity or Young's Modulus (Young's Modulus is the ratio of the compressive stress to the longitudinal strain).

　　After the yield point, the curve typically decreases slightly because of dislocations escaping from Cottrell atmospheres. As deformation continues, the stress increases on account of strain hardening until it reaches the ultimate tensile stress. Until this point, the cross-sectional area decreases uniformly and randomly because of Poisson contractions. The actual fracture point is in the same vertical line as the visual fracture point.

　　However, beyond this point a neck forms where the local cross-sectional area becomes significantly smaller than the original. If the specimen is subjected to progressively increasing tensile force it reaches the ultimate tensile stress and then necking and elongation occur rapidly until fracture. If the specimen is subjected to progressively increasing length it is possible to observe the progressive necking and elongation, and to measure the decreasing tensile force in the specimen.

　　The appearance of necking in ductile materials is associated with geometrical instability in the system. Due to the natural inhomogeneity of the material, it is common to find some regions with small inclusions or porosity within it or surface, where strain will concentrate, leading to a locally smaller area than other regions. For strain less than the ultimate tensile strain, the increase of work-hardening rate in this region will be greater than the

area reduction rate, thereby make this region harder to be further deform than others, so that the instability will be removed, i.e. the materials have abilities to weaken the inhomogeneity before reaching ultimate strain. However, as the strain become larger, the work hardening  rate will decreases, so that for now the region with smaller area is weaker than other region, therefore reduction in area will concentrate in this region and the neck becomes more and more pronounced until fracture. After the neck has formed in the materials, further plastic deformation is concentrated in the neck while the remainder of the material undergoes elastic contraction owing to the decrease in tensile force.

Brittle Materials

Brittle materials, which include cast iron, glass, and stone, are characterized by the fact that rupture occurs without any noticeable prior change in the rate of elongation.

Brittle materials such as concrete or carbon fiber do not have a yield point, and do not strain-harden. Therefore, the ultimate strength and breaking strength are the same. A typical stress strain curve is shown in Fig. 6 – 2.

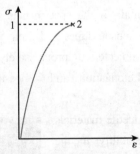

Fig. 6 – 2

Typical brittle materials like glass do not show any plastic deformation but fail while the deformation is elastic. One of the characteristics of a brittle failure is that the two broken parts can be reassembled to produce the same shape as the original component as there will not be a neck formation like in the case of ductile materials. A typical stress strain curve for a brittle material will be linear. For some materials, such as concrete, tensile strength is negligible compared to the compressive strength and it is assumed zero for many engineering applications. Glass fibers have a tensile strength stronger than steel, but bulk glass usually does not. This is because of the stress intensity factor associated with defects in the material. As the size of the sample gets larger, the size of defects also grows. In general, the tensile strength of a rope is always less than the sum of the tensile strengths of its individual fibers.

Section B Text Exploration

New Words and Expressions

analogous [ə'næləgəs]	a.	类似的，相似的，可比拟的
cavitation [ˌkævi'teiʃən]	n.	气穴现象，空穴作用，成穴
configuration [kənˌfigju'reiʃən]	n.	（各部分之间的）编排，布局；外形，轮廓；形态，构造
continuum [kən'tinjuəm]	n.	连续统一体，闭联集（复数为 continua 或 continuums）
deformation [ˌdiːfɔː'meiʃən]	n.	变形
distortion [dis'tɔːʃən]	n.	变形，失真，扭曲；曲解
dual ['djuːə]	a.	双的，双重的
	n.	双数，双数词
ductile ['dʌktail; 'dʌktil]	a.	柔软的，易延展的；易教导的
elongation [ˌiːlɔŋ'geiʃən; iˌlɔŋ'geiʃən]	n.	伸长，延长；伸长率，延伸率
fracture ['fræktʃə]	n.	破裂，断裂
	v.	使破裂，破裂，折断
inhomogeneity [inˌhɔməudʒe'niːəti]	n.	不均一，多相；不同类，不同质，不同族
Newtonian [njuː'təuniən]	a.	牛顿学说的
perpendicular [ˌpəːpən'dikjulə]	a.	垂直的，正交的，直立的，陡峭的
	n.	垂线，垂直的位置

piston	['pistən]	n.	活塞
rupture	['rʌptʃə]	n.	破裂，决裂
specimen	['spesəmən]	n.	样品，样本，标本
strain	[strein]	n.	应变；张力，拉紧，负担；扭伤；血缘
stress	[stres]	n.	压力；强调；紧张；重要性；重读
torsion	['tɔːʃən]	n.	扭转，扭曲；转矩，扭力；被扭状态
traction	['trækʃən]	n.	牵引，牵引力
truss	[trʌs]	n.	束，捆；托座，飞檐下的悬石，（支撑屋顶、桥梁等的）桁架
viscous	['viskəs]	a.	黏性的，黏的

angular distortion	角变形
compressive strain	压缩应变，压缩变形
elastic stress	弹性应力
engineering strain	工程应变
intermolecular force	分子间力，分子间作用力
macroscopic force	宏观力量
mechanical stress	机械应力，力学应力
metric tensor	度量张量，度规张量，基本张量
shearing force	剪力
shear modulus	剪切模量，剪切弹性模数
shear strain	剪切应变
shear stress	剪切应力
stress strain curve	应力应变曲线
tensile strain	拉伸应变，抗拉应变，受拉应变

Unit 6 Stress and Strain

| tensile stress | 张应力，拉伸应力 |
| viscous stress | 黏性应力 |

I True and false.

1. Stress is a physical quantity that expresses the external forces that neighboring particles of a continuous material exert on each other. (☐T ☐F)
2. Strain is the measure of the deformation of the material. (☐T ☐F)
3. Stress strain curve is unique for each material. (☐T ☐F)
4. Ductile materials are characterized by their ability to yield at normal temperatures. (☐T ☐F)
5. Brittle materials such as concrete or carbon fiber have a yield point and strain-harden. (☐T ☐F)

II Choose the best answer according to the text.

1. In some branches of engineering, the term "stress" is occasionally used in a looser sense as a synonym of _____.
   A. internal force   B. external force   C. strain   D. deformation
2. Any strain of a solid material generates an internal elastic stress, _____.
   A. which tends to change the material to another deformed state
   B. which tends to restore the material to its original non-deformed state
   C. which tends to keep the material to its deformed state like the spring
   D. which uses its reaction force to change its original state
3. A strain is in general _____.
   A. a tensor quality   B. a tensor quantity
   C. a tensile stress   D. a tensile force
4. Which does not belong to brittle materials?
   A. Cast iron.   B. Glass.   C. Stone.   D. Clay.

5. Which expression is true according to the text?
   A. If there is an increase in length of the material line, the normal strain is called compressive strain.
   B. If there is not any increase in length of the material line, the normal strain is called tensile strain.
   C. If there is reduction or compression in the length of the material line, it is called tensile strain.
   D. If there is reduction or compression in the length of the material line, it is called compressive strain.

III Translation.

1. 应变是一种变形量度，它是指物体内相对于参考长度的粒子之间的位移。变形是指物体由参考外形向当前外形的转化。
   A. Stress is a measure of deformation representing the displacement between particles in the body relative to a reference length. And strain is the shift of a body from a reference shape to a current shape.
   B. Stress is a measure of deformation representing the movement between components in the body comparing with a reference length. Deformation is the transformation of a body from a reference configuration to a current configuration.
   C. Strain is a measure of deformation representing the displacement between particles in the body relative to a reference length. Deformation is the transformation of a body from a reference configuration to a current configuration.
   D. Strain is a measure of deformation representing the movement among different components in the body comparing with a reference width. Transformation is the shift of a body from a reference shape to a current shape.

2. 典型脆性材料，例如玻璃，没有任何塑性变形，它们在弹性变形的情况下会断裂。脆性断裂的特征之一就是两个断裂的部件能够通过重组形成与原始组件相同的形状。
   A. Typical brittle materials like glass do not show any plastic deformation but fail while the deformation is elastic. One of the characteristics of a brittle failure is that the two broken parts can be reassembled to produce the same shape as the original component.
   B. Typical brittle materials like glass do not show any compressive deformation but

fail to change into elastic deformation. One of the characteristics of a brittle failure is that the two broken components can be assembled to produce the same shape as the former object.

C. Typical ductile materials like glass do not show any compressive deformation but fail to change into elastic deformation. One of the characteristics of a brittle failure is that the two broken parts can be reassembled to produce the same shape as the original component.

D. Typical ductile materials like glass do not show any plastic deformation but they are fragile while the deformation is elastic. One of the characteristics of a brittle failure is that the two broken components can be assembled to produce the same shape as the former object.

Section C Supplementary Reading

Strength of Materials

Strength of materials, also called mechanics of materials, is a subject which deals with the behavior of solid objects subject to stresses and strains. The complete theory began with the consideration of the behavior of one and two dimensional members of structures, whose states of stress can be approximated as two dimensional, and was then generalized to three dimensions to develop a more complete theory of the elastic and plastic behavior of materials. An important founding pioneer in mechanics of materials was Stephen Timoshenko.

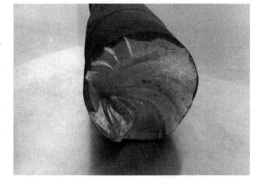

The study of strength of materials often refers to various methods of calculating the stresses and strains in structural members, such as beams, columns, and shafts. The methods employed to predict the response of a structure under loading and its susceptibility to various failure modes takes into account the properties of the materials such as its yield strength, ultimate strength, Young's modulus, and Poisson's ratio; in addition the mechanical element's macroscopic properties (geometric properties), such as its length, width, thickness, boundary constraints and abrupt changes in geometry such as holes are considered.

Definition

In materials science, the strength of a material is its ability to withstand an applied load without failure or plastic deformation. The field of strength of materials deals with forces and deformations that result from their acting on a material. A load applied to a mechanical member will induce internal forces within the member called stresses when those forces are expressed on a unit basis. The stresses acting on the material cause deformation of the material in various manners. Deformation of the material is called strain when those deformations too are placed on a unit basis. The applied loads may be axial (tensile or compressive), or rotational (strength shear). The stresses and strains that develop within a mechanical member must be calculated in order to assess the load capacity of that member. This requires a complete description of the geometry of the member, its constraints, and the loads applied to the member and the properties of the material of which the member is composed. With a complete description of the loading and the geometry of the member, the state of stress and of state of strain at any point within the member can be calculated. Once the state of stress and strain within the member is known, the strength (load carrying capacity) of that member, its deformations (stiffness qualities), and its stability (ability to maintain its original configuration) can be calculated. The calculated stresses may then be compared to some measure of the strength of the member such as its material yield or ultimate strength. The calculated deflection of the member may be compared to a deflection criterion that is based on the member's use. The calculated buckling load of the member may be compared to the applied load. The calculated stiffness and mass distribution of the member may be used to calculate the member's dynamic response and then compared to the acoustic environment in which it will be used.

Material strength refers to the point on the engineering stress-strain curve (yield stress) beyond which the material experiences deformations that will not be completely reversed upon removal of the loading and as a result the member will have a permanent deflection. The ultimate strength refers to the point on the engineering stress strain curve corresponding to the stress that produces fracture.

Types of Loadings

Transverse loading—Forces applied perpendicular to the longitudinal axis of a member. Transverse loading causes the member to bend and deflect from its original position, with internal tensile and compressive strains accompanying the change in curvature of the member. Transverse loading also induces shear forces that cause shear deformation of the material and increase the transverse deflection of the member.

Axial loading—The applied forces are collinear with the longitudinal axis of the member. The forces cause the member to either stretch or shorten.

Torsional loading—Twisting action caused by a pair of externally applied equal and oppositely directed force couples acting on parallel planes or by a single external couple applied to a member that has one end fixed against rotation.

Stress Terms

Compressive stress (or compression) is the stress state caused by an applied load that acts to reduce the length of the material (compression member) along the axis of the applied load, it is in other words a stress state that causes a squeezing of the material. A simple case

of compression is the uniaxial compression induced by the action of opposite, pushing forces. Compressive strength for materials is generally higher than their tensile strength. However, structures loaded in compression are subject to additional failure modes, such as buckling, that are dependent on the member's geometry.

Tensile stress is the stress state caused by an

applied load that tends to elongate the material along the axis of the applied load, in other words the stress caused by pulling the material. The strength of structures of equal cross-sectional area loaded in tension is independent of shape of the cross section. Materials loaded in tension are susceptible to stress concentrations such as material defects or abrupt changes in geometry. However, materials exhibiting ductile behavior (most metals for example) can tolerate some defects while brittle materials (such as ceramics) can fail well below their ultimate material strength.

Shear stress is the stress state caused by the combined energy of a pair of opposing forces acting along parallel lines of action through the material, in other words the stress caused by faces of the material sliding relative to one another. An example is cutting paper with scissors or stresses due to torsional loading.

Strength Terms

Mechanical properties of materials include the yield strength, tensile strength, fatigue strength, crack resistance, and other characteristics.

Yield strength is the lowest stress that produces a permanent deformation in a material. In some materials, like aluminums alloys, the point of yielding is difficult to identify, thus it is usually defined as the stress required causing 0.2% plastic strain. This is called a 0.2% proof stress.

Compressive strength is a limit state of compressive stress that leads to failure in a material in the manner of ductile failure (infinite theoretical yield) or brittle failure (rupture as the result of crack propagation, or sliding along a weak plane—see shear strength).

Tensile strength or ultimate tensile strength is a limit state of tensile stress that leads to tensile failure in the manner of ductile failure (yield as the first stage of that failure, some hardening in the second stage and breakage after a possible "neck" formation) or brittle failure (sudden breaking in two or more pieces at a low stress state). Tensile strength can be quoted as either true stress or engineering stress, but engineering stress is the most commonly used.

Fatigue strength is a measure of the strength of a material or a component under cyclic loading, and is usually more difficult to assess than the static strength measures. Fatigue strength is quoted as stress amplitude or stress range, usually at zero means stress, along with the number of cycles to failure under that condition of stress.

Impact strength, is the capability of the material to withstand a suddenly applied load and is expressed in terms of energy. Often measured with the Izod impact strength test or Charpy impact test, both of which measure the impact energy required to fracture a sample. Volume, modulus of elasticity, distribution of forces, and yield strength affect the impact strength of a material. In order for a material or object to have high impact strength the stresses must be distributed evenly throughout the object. It also must have a large volume with a low modulus of elasticity and high material yield strength.

Strain (Deformation) Terms

Deformation of the material is the change in geometry created when stress is applied (as a result of applied forces, gravitational fields, accelerations, thermal expansion, etc.). Deformation is expressed by the displacement field of the material.

Strain or reduced deformation is a mathematical term that expresses the trend of the deformation change among the material field. Strain is the deformation per unit length. In the case of uniaxial loading the displacements of a specimen (for example a bar element) lead to a calculation of strain expressed as the quotient of the displacement and the original length of the specimen. For 3D displacement fields it is expressed as derivatives of displacement functions in terms of a second order tensor (with 6 independent elements).

Deflection is a term to describe the magnitude to which a structural element is displaced when subject to an applied load.

Stress-strain Relations

Elasticity is the ability of a material to return to its previous shape after stress is released. In many materials, the relation between applied stress is directly proportional to the resulting strain (up to a certain limit), and a graph representing those two quantities is a straight line.

The slope of this line is known as Young's modulus, or the "modulus of elasticity". The modulus of elasticity can be used to determine the stress-strain relationship in the linear-elastic portion of the stress-strain curve. The linear-elastic region is either below the yield point, or if a yield point is not easily identified on the stress-strain plot it is defined to be between 0 and 0.2% strain, and is defined as the region of strain in which no yielding (permanent deformation) occurs.

Plasticity or plastic deformation is the opposite of elastic deformation and is defined as unrecoverable strain. Plastic deformation is retained after the release of the applied stress. Most materials in the linear-elastic category are usually capable of plastic deformation. Brittle materials, like ceramics, do not experience any plastic deformation and will fracture under relatively low strain, while ductile materials such as metallic, lead, or polymers will plastically deform much more before a fracture initiation.

Consider the difference between a carrot and chewed bubble gum. The carrot will stretch very little before breaking. The chewed bubble gum, on the other hand, will plastically deform enormously before finally breaking.

Design Terms

Ultimate strength is an attribute related to a material, rather than just a specific specimen made of the material, and as such it is quoted as the force per unit of cross section area (N/m^2). The ultimate strength is the maximum stress that a material can withstand before it breaks or weakens. For example, the ultimate tensile strength (UTS)

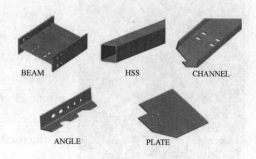

BEAM　　HSS　　CHANNEL

ANGLE　　PLATE

of AISI 1018 Steel is 440 MN/m². In general, the SI unit of stress is the pascal, where 1 Pa = 1 N/m².

Design stresses that have been determined from the ultimate or yield point values of the materials give safe and reliable results only for the case of static loading. Many machine parts fail when subjected to non-steady and continuously varying loads even though the developed stresses are below the yield point. Such failures are called fatigue failure. The failure is by a fracture that appears to be brittle with little or no visible evidence of yielding. However, when the stress is kept below "fatigue stress" or "endurance limit stress", the part will endure indefinitely. A purely reversing or cyclic stress is one that alternates between equal positive and negative peak stresses during each cycle of operation. In a purely cyclic stress, the  average stress is zero. When a part is subjected to a cyclic stress, also known as stress range (Sr), it has been observed that the failure of the part occurs after a number of stress reversals (N) even if the magnitude of the stress range is below the material's yield strength. Generally, higher the range stress, the fewer the number of reversals needed for failure.

Failure Theories

There are four failure theories: maximum shear stress theory, maximum normal stress theory, maximum strain energy theory, and maximum distortion energy theory. Out of these four theories of failure, the maximum normal stress theory is only applicable for brittle materials, and the remaining three theories are applicable for ductile materials. Of the latter three, the distortion energy theory provides most accurate results in majority of the stress conditions. The strain energy theory needs the value of Poisson's ratio of the part material, which is often not readily available. The maximum shear stress theory is conservative. For simple unidirectional normal stresses all theories are equivalent, which means all theories will give the same result.

A material's strength is dependent on its microstructure. The engineering processes to which a material is subjected can alter this microstructure. The variety of strengthening mechanisms that alter the strength of a material includes work hardening, solid solution strengthening, precipitation hardening and grain boundary strengthening and can be quantitatively and qualitatively explained. Strengthening mechanisms are accompanied by the caveat that some other mechanical properties of the material may degenerate in an attempt to make the material stronger. For example, in grain boundary strengthening, although yield strength is maximized with decreasing grain size, ultimately, very small grain sizes make the material brittle. In general, the yield strength of a material is an adequate indicator of the material's mechanical strength. Considered in tandem with the fact that the yield strength is the parameter that predicts plastic deformation in the material, one can make informed decisions on how to increase the strength of a material depending on its microstructural properties and the desired end effect. Strength is expressed in terms of the limiting values of the compressive stress, tensile stress, and shear stresses that would cause failure. The effects of dynamic loading are probably the most important practical consideration of the strength of materials, especially the problem of fatigue. Repeated loading often initiates brittle cracks, which grow until failure occurs. The cracks always start at stress concentrations, especially changes in cross-section of the product, near holes and corners at nominal stress levels far lower than those quoted for the strength of the material.

Unit 7
Surveying

Section A Text

Surveying

Surveying or land surveying is the technique, profession, and science of determining the terrestrial or three-dimensional position of points and the distances and angles between them. A land surveying professional is called a land surveyor. These points are usually on the surface of the Earth, and they are often used to establish land maps and boundaries for ownership, locations like building corners or the surface location of subsurface features, or other purposes required by government or civil law, such as property sales. Surveying has been an element in the development of the human environment since the beginning of recorded history. The planning and execution of most forms of construction require it. It is an important tool for research in many other scientific disciplines.

Basic surveying has occurred since humans built the first large structures. In ancient Egypt, a rope stretcher would use simple geometry to re-establish boundaries after the annual floods of the Nile River. In England, William the Conqueror commissioned the Domesday Book in 1086. It recorded the names of all the land owners, the area of land they owned, the quality of the land, and specific information of the area's content and inhabitants.

As for modern surveying, Gunter's chain was introduced in 1620 by English mathematician Edmund Gunter. It enabled plots of land to be accurately surveyed and plotted for legal and commercial purposes. In the 18th century, modern techniques and instruments for surveying began to be used. Jesse Ramsden introduced the first precision theodolite in 1787. It was an instrument for measuring angles in the horizontal and vertical planes. He created his great theodolite using an accurate dividing engine of his own design. Ramsden's theodolite represented a great step forward in the instrument's accuracy. Between 1733 and 1740, Jacques Cassini and his son César undertook the first triangulation of France. They included a re-surveying of the meridian arc, leading to the publication in 1745 of the first map of France constructed on rigorous principles. By this time, triangulation methods were by then well established for local map-making. Surveying became a professional occupation in high demand at the turn of the 19th century with the onset of the Industrial Revolution. The profession developed more accurate instruments to aid its work. In the US, the Land Ordinance of 1785 created the Public Land Survey System (PLSS). It formed the basis for dividing the western territories into sections to allow the sale of land. The PLSS divided states into township grids which were further divided into sections and fractions of sections. Napoleon Bonaparte founded continental Europe's first cadastre in 1808. This gathered data on the number of parcels of land, their value, land usage, and names. This system soon spread around Europe.

At the beginning of 20th century surveyors had improved the older chains and ropes, but still faced the problem of accurate measurement of long distances. Advances in electronics allowed miniaturization of electronic distance measurement (EDM). In the 1970s the first instruments combining angle and distance measurement appeared, becoming known as total stations. Manufacturers added more equipment by degrees, bringing improvements in accuracy and speed of measurement. Major advances include tilt compensators, data recorders, and on-board calculation programs. The US Air Force

launched the first prototype satellites of the Global Positioning System (GPS) in 1978. GPS used a larger constellation of satellites and improved signal transmission to provide more accuracy. Early GPS observations required several hours of observations by a static receiver to reach survey accuracy requirements. Recent improvements to both satellites and receivers allow real time kinematic (RTK) surveying. RTK surveys get high-accuracy measurements by using a fixed base station and a second roving antenna. The position of the roving antenna can be tracked.

In the 21st century, the theodolite, total station, and RTK, GPS survey remain the primary methods in use. Remote sensing and satellite imagery continue to improve and become cheaper, allowing more commonplace use. Prominent new technologies include three-dimensional (3D) scanning and use of lidar for topographical surveys. Unmanned aerial vehicle (UAV) technology along with photogrammetric image processing is also appearing.

The main surveying instruments in use around the world are the theodolite, measuring tape, total station, 3D scanners, GPS/GNSS (Global Navigation Satellite System), level and rod. Most instruments screw onto a tripod when in use. Tape measures are often used for measurement of smaller distances. 3D scanners and various forms of aerial imagery are applied. In addition, some surveying techniques are required, including distance measurement, angle measurement, leveling, determining position, triangulation, traversing, datum and coordinate systems.

A basic tenet of surveying is that no measurement is perfect, and that there will always be a small amount of error. There are three classes of survey errors.

(1) Gross errors or blunders: Errors made by the surveyor during the survey. Upsetting the instrument, misaiming a target, or writing down a wrong measurement are all gross errors. A large gross error may reduce the accuracy to an unacceptable level. Therefore, surveyors use redundant measurements and independent checks to detect these errors early in the survey.

(2) Systematic: Errors that follow a consistent pattern. Examples include effects of temperature on a chain or EDM measurement, or a poorly adjusted spirit level causing a tilted

instrument or target pole. Systematic errors that have known effects can be compensated or corrected.

(3) Random: Random errors are small unavoidable fluctuations. They are caused by imperfections in measuring equipment, eyesight, and conditions. They can be minimized by redundancy of measurement and avoiding unstable conditions. Random errors tend to cancel each other out, but checks must be made to ensure they are not propagating from one measurement to the next.

Surveyors avoid these errors by calibrating their equipment, using consistent methods, and by good design of their reference network. Repeated measurements can be averaged and any outlier measurements discarded. Independent checks like measuring a point from two or more locations or using two different methods are used. Errors can be detected by comparing the results of the two measurements.

Local organizations or regulatory bodies class specializations of surveying in different ways. The main types of surveys are as follows:

(1) Cadastral or boundary survey: a survey that establishes or re-establishes boundaries of a parcel using a legal description. It involves the setting or restoration of monuments or markers at the corners or along the lines of the parcel. These take the form of iron rods, pipes, or concrete monuments in the ground, or nails set in concrete or asphalt. It incorporates elements of the boundary survey, mortgage survey, and topographic survey.

(2) Control survey: Control surveys establish reference points to use as starting positions for future surveys. Most other forms of surveying will contain elements of control surveying.

(3) Construction survey: Construction surveying is generally performed by specialized technicians. Unlike land surveyors, the resulting plan does not have legal status. Construction surveyors perform the following tasks: Surveying existing conditions of the future work site, including topography, existing buildings and infrastructure, and underground infrastructure when possible.

(4) Deformation survey: a survey to determine if a structure or object is changing shape or moving.

First the positions of points on an object are found. A period of time is allowed to pass and the positions are then re-measured and calculated. Then a comparison between the two sets of positions is made.

(5) Dimensional control survey: This is a type of survey conducted in or on a non-level surface. Common in the oil and gas industry to replace old or damaged pipes on a like-for-like basis, the advantage of dimensional control survey is that the instrument used to conduct the survey does not need to be level. This is useful in the off-shore industry, as not all platforms are fixed and are thus subject to movement.

(6) Foundation survey: a survey done to collect the positional data on a foundation that has been poured and is cured. This is done to ensure that the foundation was constructed in the location, and at the elevation, authorized in the plot plan, site plan, or subdivision plan.

(7) Hydrographic survey: a survey conducted with the purpose of mapping the shoreline and bed of a body of water. It's used for navigation, engineering, or resource management purposes.

(8) Leveling: either finds the elevation of a given point or establish a point at a given elevation.

(9) Mining survey: Mining surveying includes directing the digging of mine shafts and galleries and the calculation of volume of rock. It uses specialized techniques due to the restraints to survey geometry such as vertical shafts and narrow passages.

(10) Photographic control survey: a survey that creates reference marks visible from the air to allow aerial photographs to be rectified.

(11) Structural survey: a detailed inspection to report upon the physical condition and structural stability of a building or structure. It highlights any work needed to maintain it in good repair.

(12) Subdivision: A boundary survey that splits a property into two or more smaller

properties.

(13) Topographic survey: a survey that measures the elevation of points on a particular piece of land, and presents them as contour lines on a plot.

The basic principles of surveying have changed little over the ages, but the tools used by surveyors have evolved. Engineering, especially civil engineering, often needs surveyors. Surveyors help determine the placement of roads, railways, reservoirs, dams, pipelines, retaining walls, bridges, and buildings. They establish the boundaries of legal descriptions and political divisions. They also provide advice and data for Geographical Information Systems (GIS) that record land features and boundaries. Surveyors must have a thorough knowledge of algebra, basic calculus, geometry, and trigonometry. They must also know the laws that deal with surveys, real property, and contracts.

Most jurisdictions recognize three different levels of qualification.

(1) Survey assistants or chainmen are usually unskilled workers who help the surveyor. They place target reflectors, find old reference marks, and mark points on the ground. The term "chainman" derives from past use of measuring chains. An assistant would move the far end of the chain under the surveyor's direction.

(2) Survey technicians often operate survey instruments, run surveys in the field, do survey calculations, or draft plans. A technician usually has no legal authority and cannot certify his work. Not all technicians are qualified, but qualifications at the certificate or diploma level are available.

(3) Licensed, registered, or chartered surveyors usually hold a degree or higher qualification. They are often required to pass further exams to join a professional association or to gain certifying status. Surveyors are responsible for planning and management of surveys. They have to ensure that their surveys, or surveys performed under their supervision, meet the legal standards. Many principals of surveying firms hold this status.

Most jurisdictions also have a form of professional institution representing local surveyors. These institutes often endorse or license potential surveyors, as well as set and enforce ethical standards. The largest institution is the International Federation of Surveyors (FIG). They represent the survey industry worldwide.

Section B Text Exploration

New Words and Expressions

aerial ['ɛəriəl]	a.	航空的，飞机的，飞行的，空中的
algebra ['ældʒibrə]	n.	代数（学）
asphalt ['æsfælt]	n.	沥青，柏油
	a.	用柏油铺成的
blunder ['blʌndə]	v.	跌跌撞撞地走；犯大错，做错
	n.	大错
cadastre [kə'dæstə]	n.	地籍簿
calibrate ['kælibreit]	v.	校正，调整，测定口径
chainman ['tʃeinmən]	n.	测链员，丈量员，测量时拿皮尺或卷尺的人
compensator ['kɔmpənseitə]	n.	补偿器，自耦变压器；赔偿者；补偿物
constellation [ˌkɔnstə'leiʃən]	n.	星座，星群；荟萃
datum ['deitəm]	n.	数据，资料；基点，基线，基面；已知数
endorse [in'dɔːs]	v.	认可，赞同；签署，在背面签名
fluctuation [ˌflʌktju'eiʃən]	n.	（状态、方向、位置等）不稳定，不断变化，波动，起伏
geometry [dʒi'ɔmitri]	n.	几何学，几何结构
grid [grid]	n.	网格，格子，栅格；输电网

jurisdiction [ˌdʒuəris'dikʃən]	n.	司法权，审判权，管辖权；权限，权力
leveling ['levəliŋ]	n.	水准测量
lidar ['laidɑː]	n.	激光雷达，激光定位器
miniaturization [ˌminiətʃərai'zeiʃən]	n.	小型化，微型化
outlier ['autˌlaiə]	n.	异常值；露宿者；局外人；离开本体的部分
photogrammetric [ˌfəutəugrə'metrik]	a.	摄影制图的，摄影测量的
propagate ['prɔpəgeit]	v.	传播，传送；繁殖，增殖
prototype ['prəutətaip]	n.	原型，标准，模范
redundant [ri'dʌndənt]	a.	多余的，过剩的；被解雇的，失业的；冗长的，累赘的
screw [skruː]	v.	转动，旋，拧；压榨，强迫
	n.	螺旋，螺丝钉
tenet ['tiːnet]	n.	原则，信条，教义
terrestrial [ti'restriəl]	a.	地球的，陆地的
	n.	陆地生物，地球上的人
topographical [ˌtɔpə'græfikəl]	a.	地质的，地形学的
triangulation [traiˌæŋgju'leiʃən]	n.	三角测量，三角形划分
trigonometry [ˌtrigə'nɔmitri]	n.	三角法
tripod ['traipɔd]	n.	三脚架，三脚桌

base station	基站，基电台
construction survey	施工测量，建筑测量
contour line	轮廓线，等高线
coordinate system	坐标系
dividing engine	刻度机，分度器
Gunter's chain	冈特测链，冈氏测链
hydrographic survey	水文测量，水道测量
measuring tape	卷尺，皮尺
meridian arc	子午线弧

Unit 7　Surveying

real time kinematic（RTK）	实时动态
spirit level	水平尺，水准仪
topographic survey	地形测量，地形勘察
unmanned aerial vehicle（UAV）	无人驾驶飞行器

Domesday Book　　　　　　　　　　　英国国王1086年颁布的土地志
Geographical Information Systems（GIS）　地理信息系统
International Federation of Surveyors（FIG）　国际测量员联合会
Public Land Survey System（PLSS）　　公共土地测量系统
William the Conqueror　　　　　　　英国国王威廉一世

I True and false.

1. Surveying or land surveying is only applied in the field of construction. (☐T　☐F)
2. A basis tenet of surveying is that no measurement is perfect and that there will always be a small amount of error. (☐T　☐F)
3. Basic surveying has occurred since the Industrial Revolution. (☐T　☐F)
4. Foundation survey measures the elevation of points on a particular piece of land, and presents them as contour lines on a plot. (☐T　☐F)
5. Most jurisdictions have a form of professional institution representing local surveyors and the largest one is FIG. (☐T　☐F)

II Choose the best answer according to the text.

1. Which of the following does not belong to the achievements in modern surveying?
   A. Jesse Ramsden introduced the first precision theodolite.
   B. Jacques Cassini and his son César undertook the first triangulation of France.
   C. William the Conqueror commissioned the Domesday Book.
   D. Edmund Gunter introduced Gunter's chain.

2. The main surveying instruments in use around the world are _____.
   A. the theodolite, measuring tape and total station
   B. level and rod
   C. 3D scanners and GPS/GNSS
   D. all of A, B and C

3. _____ is a survey to determine if a structure or object is changing shape or moving.
   A. Control survey   B. Leveling
   C. Deformation survey   D. Mining survey

4. As for survey errors, which statement is not true according to the text?
   A. A large gross error may reduce the accuracy to an unacceptable level.
   B. Systematic errors that have known effects cannot be compensated or corrected.
   C. Upsetting the instrument, misaiming a target, or writing down a wrong measurement are all gross errors.
   D. Blunders are errors made by the surveyor during the survey.

5. Survey technicians _____.
   A. often operate survey instruments, run surveys in the field, do survey calculations, or draft plans
   B. usually hold a degree or higher qualification
   C. are usually unskilled workers who help the surveyor
   D. can also be called chartered surveyors

III Translation.

1. 随机误差是微小的不可避免的波动。它们是由于测量设备、视力和条件的不完善造成的。

A. Random errors are small unavoidable fluctuations. They are caused by imperfections in measuring equipment, eyesight, and conditions.

B. Random errors are small avoidable fluctuations. They give rise to imperfections in measuring equipment, eyesight, and conditions.

C. Gross errors are small avoidable fluctuations. They are caused by imperfections in measuring equipment, eyesight, and conditions.

D. Gross errors are small unavoidable fluctuations. They give rise to imperfections in measuring equipment, eyesight, and conditions.

2. 测量或陆地测量是一门确定点的地面位置或三维位置及点间距离和角度的技术、职业和科学。

A. Surveying or building surveying is the technique, profession, and science of determining the terrestrial or three-dimensional position of points and the length and angles between them.

B. Surveying or land surveying is the technique, profession, and discipline of determining the terrestrial or three-dimensional position of points and the distances and angles between them.

C. Surveying or building surveying is the technique, profession, and discipline of determining the terrestrial or three-dimensional position of points and the distances and angles between them.

D. Surveying or land surveying is the technique, profession, and science of determining the terrestrial or three-dimensional position of points and the distances and angles between them.

3. 矿井测量包括指引挖掘矿井和通道的方位，以及计算岩石的体积。

A. Leveling surveying includes directing the digging of mine shafts and galleries and the calculation of volume of pebble.

B. Mining surveying includes directing the digging of mine shafts and galleries and the calculation of volume of rock.

C. Leveling surveying includes directing the digging of mine shafts and galleries and the calculation of weight of rock.

D. Mining surveying includes directing the digging of mine shafts and galleries and the calculation of volume of pebble.

Section C Supplementary Reading

3D Scanner

A 3D scanner is a device that analyses a real-world object or environment to collect data on its shape and possibly its appearance (e.g. color). The collected data can then be used to construct digital three-dimensional models. Many different technologies can be used to build these 3D-scanning devices; each technology comes with its own limitations, advantages and costs. Many limitations in the kind of objects that can be digitized are still present, for example, optical technologies encounter many difficulties with shiny, mirroring or transparent objects. For example, industrial computed tomography scanning can be used to construct digital 3D models, applying non-destructive testing. Collected 3D data is useful for a wide variety of applications. These devices are used extensively by the entertainment industry in the production of movies and video games. Other common applications of this technology include industrial design, orthotics and prosthetics, reverse engineering and prototyping, quality control/inspection and documentation of cultural artifacts.

Functionality

The purpose of a 3D scanner is usually to create a point cloud of geometric samples on the surface of the subject. These points can then be used to extrapolate the shape of the subject (a process called reconstruction). If color information is collected at each point, then the colors on the surface of the subject can also be determined.

3D scanners share several traits with cameras. Like most cameras, they have a cone-like field of view, and like cameras, they can only collect information about surfaces that are not obscured. While a camera collects color information about surfaces within its field of view, a 3D scanner collects distance information about surfaces within its field of view. The "picture" produced by a 3D scanner describes the distance to a surface at each point in the picture. This allows the three dimensional position of each point in the picture to be identified.

Reconstruction

From point clouds: The point clouds produced by 3D scanners and 3D imaging can be used directly for measurement and visualization in the architecture and construction world.

From models: Most applications, however, use instead polygonal 3D models, Non-Uniform Rational B-Splines (NURBS) surface models, or editable feature-based CAD models.

(1) Polygon mesh models. In a polygonal representation of a shape, a curved surface is modeled as many small faceted flat surfaces (think of a sphere modeled as a disco ball). Reconstruction to polygonal model involves finding and connecting adjacent points with straight lines in order to create a continuous surface.

(2) Surface models. The next level of sophistication in modeling involves using a quilt of curved surface patches to model the shape. These might be NURBS, TSplines or other curved representations of curved topology. Using NURBS, the spherical shape becomes a true mathematical sphere. Some applications offer patch layout by hand but the best in class offer both automated patch layout and manual layout. These patches have the advantage of being lighter and more manipulable when exported to CAD. Surface models are somewhat editable, but only in a sculptural sense of pushing and pulling to deform the surface.

(3) Solid CAD models. From an engineering/manufacturing perspective, the ultimate representation of a digitized shape is the editable, parametric CAD model. In CAD, the sphere

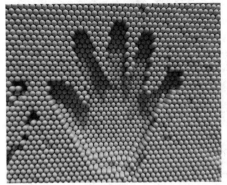

is described by parametric features which are easily edited by changing a value (e.g., centre point and radius). These CAD models describe not simply the envelope or shape of the object, but CAD models also embody the "design intent" (i.e., critical features and their relationship to other features). An example of design intent not evident in the shape alone might be a brake drum's lug bolts, which must be concentric with the hole in the centre of the drum. This knowledge would drive the sequence and method of creating the CAD model; a designer with an awareness of this relationship would not design the lug bolts referenced to the outside diameter, but instead, to the center. A modeler creating a CAD model will want to include both Shape and design intent in the complete CAD model.

From a set of 2D slices: CT, industrial CT, MRI, or Micro-CT scanners do not produce point clouds but a set of 2D slices which are then "stacked together" to produce a 3D representation. There are several ways to do this depending on the output required.

(1) Volume rendering: Different parts of an object usually have different threshold values or greyscale densities. From this, a 3-dimensional model can be constructed and displayed on screen. Multiple models can be constructed from various thresholds, allowing different colors to represent each component of the object. Volume rendering is usually only used for visualization of the scanned object.

(2) Image segmentation: Where different structures have similar threshold values, it can become impossible to separate them simply by adjusting volume rendering parameters. The solution is called segmentation, a manual or automatic procedure that can remove the unwanted structures from the image. Image segmentation software usually allows export of the segmented structures in CAD or STL format for further manipulation.

(3) Image-based meshing: When using 3D image data for computational analysis (e.g. CFD and FEA), simply segmenting the data and meshing from CAD can become time consuming, and virtually intractable for the complex topologies typical of image data. The solution is called image-based meshing, an automated process of generating an accurate and realistic geometrical description of the scan data.

From laser scans: Laser scanning describes the general method to sample or scan a surface using laser technology. Several areas of application exist that mainly differ in the power of the lasers that are used, and in the results of the scanning process. Low laser power is used when the scanned surface doesn't have to be influenced. Confocal or 3D laser scanning are methods to get information about the scanned surface. Another low-power application uses structured light projection systems for solar cell flatness metrology, enabling stress calculation throughout in excess of 2000 wafers per hour.

Applications

1. Construction industry and civil engineering

As-built drawings of bridges, industrial plants, and monuments

Documentation of historical sites

Site modeling and lay outing

Quality control

Quantity surveys

Payload monitoring

Freeway redesign

Forensic documentation

2. Design process

Increasing accuracy working with complex parts and shapes, coordinating product design using parts from multiple sources, updating old CD scans with those from more current technology, replacing missing or older parts, creating cost savings by allowing as-built design services, for example in automotive manufacturing plants, and saving travel costs.

3. Entertainment

3D scanners are used by the entertainment industry to create digital 3D models for movies, video games and leisure purposes. They are heavily utilized in virtual cinematography. In cases where a real-world equivalent of a model exists, it is much faster to scan the real-world object than to manually create a model using 3D modeling software. Frequently, artists sculpt physical models of what they want and scan them into digital form rather than directly creating digital models on a computer.

4. 3D photography

3D scanners are evolving for the use of cameras to represent 3D objects in an accurate manner. Companies are emerging since 2010 that create 3D portraits of people (3D figurines).

5. Law enforcement

3D laser scanning is used by the law enforcement agencies around the world. 3D Models are used for on-site documentation of crime scenes, bullet trajectories, bloodstain pattern analysis, accident reconstruction, bombings, plane crashes, and more.

6. Reverse engineering

Reverse engineering of a mechanical component requires a precise digital model of the objects to be reproduced. Rather than a set of points a precise digital model can be represented by a polygon mesh, a set of flat or curved NURBS surfaces, or ideally for mechanical components, a CAD solid model. A 3D scanner can be used to digitize free-form or gradually changing shaped components as well as prismatic geometries whereas a coordinate measuring machine is usually used only to determine simple dimensions of a highly prismatic model. These data points are then processed to create a usable digital model, usually using specialized reverse engineering software.

7. Real estate

Land or buildings can be scanned into a 3d model, which allows buyers to tour and inspect the property remotely, anywhere, without having to be present at the property. There is already at least one company providing 3d-scanned virtual real estate tours. A typical virtual tour would consist of dollhouse view, inside view, as well as a floor plan.

8. Virtual/Remote tourism

The environment at a place of interest can be captured and converted into a 3D model. This model can then be explored by the public, either through a VR interface or a traditional "2D" interface. This allows the user to explore locations which are inconvenient for travel.

9. Cultural heritage

There have been many research projects undertaken via the scanning of historical sites and artifacts both for documentation and analysis purposes.

The combined use of 3D scanning and 3D printing technologies allows the replication of real objects without the use of traditional plaster casting techniques, that in many cases can be too invasive for being performed on precious or delicate cultural heritage artifacts. In an example of a typical application scenario, a gargoyle model was digitally acquired using a 3D scanner and the produced 3D data was processed using MeshLab. The resulting digital 3D model was fed to a rapid prototyping machine to create a real resin replica of the original object.

10. Medical CAD/CAM

3D scanners are used to capture the 3D shape of a patient in orthotics and dentistry. It gradually supplants tedious plaster cast. CAD/CAM software is then used to design and manufacture the orthosis, prosthesis or dental implants.

Many Chairside dental CAD/CAM systems and Dental Laboratory CAD/CAM systems use 3D Scanner technologies to capture the 3D surface of a dental preparation, in order to produce a restoration digitally using CAD software and ultimately produce the final restoration using a CAM technology (such as a CNC milling machine, or 3D printer). The chairside systems are designed to facilitate the 3D scanning of a preparation in vivo and produce the restoration.

11. Quality assurance and industrial metrology

The digitalization of real-world objects is of vital importance in various application domains. This method is especially applied in industrial quality assurance to measure the geometric dimension accuracy. Industrial processes such as assembly are complex, highly automated and typically based on CAD (Computer Aided Design) data. The problem is that the same degree of automation is also required for quality assurance. It is, for example, a very complex task to assemble a modern car, since it consists of many parts that must fit together at the very end of the production line. The optimal performance of this process is guaranteed by quality assurance systems. Especially the geometry of the metal parts must be checked in order to assure that they have the correct dimensions, fit together and finally work reliably.

Within highly automated processes, the resulting geometric measures are transferred to machines that manufacture the desired objects. Due to mechanical uncertainties and abrasions, the result may differ from its digital nominal. In order to automatically capture and evaluate these deviations, the manufactured part must be digitized as well. For this purpose, 3D scanners are applied to generate point samples from the object's surface which are finally compared against the nominal data.

The process of comparing 3D data against a CAD model is referred to as CAD-Compare, and can be a useful technique for applications such as determining wear patterns on moulds and tooling, determining accuracy of final build, analyzing gap and flush, or analyzing highly complex sculpted surfaces. At present, laser triangulation scanners, structured light and contact scanning are the predominant technologies employed for industrial purposes, with contact scanning remaining the slowest, but overall most accurate option. Nevertheless, 3D scanning technology offers distinct advantages compared to traditional touch probe measurements. White-light or laser scanners accurately digitize objects all around, capturing fine details and freeform surfaces without reference points or spray. The entire surface is covered at record speed without the risk of damaging the part. Graphic comparison charts illustrate geometric deviations of full object level, providing deeper insights into potential causes.

Unit 8
Tall Buildings

Section A Text

Tall Buildings

A tall building or a high-rise building is a multi-story structure between 35-100 meters tall, or a building of unknown height from 12-39 floors and a skyscraper is a multi-story building whose architectural height is at least 100 meters or 330 feet. The word skyscraper often carries a connotation of pride and achievement. Some structural engineers define a hig-hrise as any vertical construction for which wind is a more significant load factor than earthquake or weight.

"Vanity height" is the distance between the highest floor and its architectural top (excluding antennae, flagpole or other functional extensions). The current era of tall buildings focuses on sustainability, its built and natural environments, including the performance of structures, types of materials, construction practices, absolute minimal use of materials and natural resources, energy  within the structure, and a holistically integrated building systems approach.

Design and Construction

The design and construction of tall buildings involves creating safe, habitable spaces in very tall buildings. The buildings must support their weight, resist wind and earthquakes, and protect occupants from fire. Yet they must also be conveniently accessible, even on the upper floors, and provide utilities and a comfortable climate for the occupants. The problems posed in a tall building design are considered among the most complex encountered given the balances required between economics, engineering, and construction management.

One common feature of tall buildings is a steel framework from which curtain walls are suspended, rather than load-bearing walls of conventional construction. Most tall buildings have a steel frame that enables them to be built taller than typical load-bearing walls of reinforced concrete. Tall buildings usually have a particularly small surface area of what are conventionally thought of as walls. Because the walls are not load-bearing most tall buildings are characterized by surface areas of windows made possible by the concept of steel frame and curtain wall. However, tall buildings can also have curtain walls that mimic conventional walls and have a small surface area of windows.

The construction of tall buildings was enabled by steel frame construction that surpassed brick and mortar construction starting at the end of the 19th century and finally surpassing it in the 20th century together with reinforced concrete construction as the price of steel decreased and labor costs increased. The steel frames become inefficient and uneconomic for supertall buildings as usable floor space is reduced for progressively larger supporting columns. Since about 1960, tubular designs have been used for high rises. This reduces the usage of material yet allows greater height. It allows fewer interior columns, and so creates more usable floor space. It further enables buildings to take on various shapes.

Elevators are characteristic to tall buildings. In 1852 Elisha Otis introduced the safety elevator, allowing convenient and safe passenger movement to upper floors. Another crucial development was the use of a steel frame instead of stone or brick, otherwise the walls on the lower floors on a tall building would be too thick to be practical. Advances in construction techniques have allowed tall buildings to narrow in width, while increasing in height. Some of these new techniques include mass dampers to reduce vibrations and swaying, and gaps to allow air to pass through, reducing wind shear.

Loading and Vibration

The load a tall building experiences is largely from the force of the building material itself. In most building designs, the weight of the structure is much larger than the weight of the material that it will support beyond its own weight. In technical terms, the dead load, the load of the structure, is larger than the live load, the weight of things in the structure

(people, furniture, vehicles, etc.). As such, the amount of structural material required within the lower levels of a tall building will be much larger than the material required within higher levels. This is not always visually apparent. Vertical supports can come in several types, among which the most common for tall buildings can be categorized as steel frames, concrete cores, tube within tube design, and shear walls.

The wind loading on a tall building is also considerable. In fact, the lateral wind load imposed on super-tall structures is generally the governing factor in the structural design. Wind pressure increases with height, so for very tall buildings, the loads associated with wind are larger than dead or live loads. Other vertical and horizontal loading factors come from varied, unpredictable sources, such as earthquakes.

Steel Frame

By 1895, steel had replaced cast iron as tall buildings' structural material. Its malleability allowed it to be formed into a variety of shapes, and it could be riveted, ensuring strong connections. The simplicity of a steel frame eliminated the inefficient part of a shear wall, the central portion, and consolidated support members in a much stronger fashion by allowing both

horizontal and vertical supports throughout. Among steel's drawbacks is that as more material must be supported as height increases, the distance between supporting members must decrease, which in turn increases the amount of material that must be supported. This becomes inefficient and uneconomic for buildings above 40 stories tall as usable floor spaces are reduced for supporting column and due to more usage of steel.

Tube Structural Systems

A new structural system of framed tubes was developed in 1963. Fazlur Khan and J. Rankine defined the framed tube structure as a three dimensional space structure composed of three, four, or possibly more frames, braced frames, or shear walls, joined at or near their edges to form a vertical tube-like structural system capable of resisting lateral forces in any direction by cantilevering from the foundation. Closely spaced interconnected exterior columns form the tube. Horizontal loads (primarily wind) are supported by the structure as a whole. Framed tubes allow fewer interior columns, and so create more usable floor space, and about half the exterior surface is available for windows. Where larger openings like garage doors are required, the tube frame must be interrupted, with transfer girders used to maintain structural integrity. Tube structures cut down costs, at the same time allowing buildings to reach greater heights. The tubular systems are fundamental to tall building design.

Trussed Tube and X-bracing

Khan pioneered several other variations of the tube structure design. One of these was the concept of X-bracing, or the "trussed tube", first employed for the John Hancock Center. This concept reduced the lateral load on the building by transferring the load into the exterior columns. This allows for a reduced need for interior columns thus creating more floor space.

This concept can be seen in the John Hancock Center, designed in 1965 and completed in

1969. One of the most famous buildings of the structural expressionist style, the tall building's distinctive X-bracing exterior is actually a hint that the structure's skin is indeed part of its "tubular system". This idea is one of the architectural techniques the building used to climb to record heights (the tubular system is essentially the spine that helps the building stand upright during wind and earthquake loads). This X-bracing allows for both higher performance from tall structures and the ability to open up the inside floorplan if the architect desires.

Bundled Tube

An important variation on the tube frame is the "bundled tube", which uses several interconnected tube frames. The Willis Tower in Chicago used this design, employing nine tubes of varying height to achieve its distinct appearance. The bundled tube structure meant that buildings no longer needed to be boxlike in appearance: they could become sculpture.

The Elevator Conundrum

The invention of the elevator was a precondition for the invention of tall buildings, given that most people would not (or could not) climb more than a few flights of stairs at a time. The elevators in a tall building are not simply a necessary utility, like running water and electricity, but are in fact closely related to the design of the whole structure: a taller building requires more elevators to service the additional floors, but the elevator shafts consume valuable floor space. If the service core, which contains the elevator shafts, becomes too big, it can reduce the profitability of the building. Architects must therefore balance the value gained by adding height against the value lost to the expanding service core.

Many tall buildings use elevators in a non-standard configuration to reduce their footprint. Buildings such as the former World Trade Center Towers and Chicago's John Hancock Center use sky lobbies, where express elevators take passengers to upper floors which serve as the base for local elevators. This allows architects and engineers to place elevator shafts on top of each other, saving space. Sky lobbies and express elevators take up a significant amount of space, however, and add to the amount of time spent commuting between floors.

Economic Rationale

Tall buildings are usually situated in city centers where the price of land is high. Constructing a tall building becomes justified if the price of land is so high that it makes economic sense to build upwards as to minimize the cost of the land per the total floor area of a building. Thus the construction of tall buildings is dictated by economics and results in tall buildings in a certain part of a large city unless a building code restricts the height of buildings.

Tall buildings are rarely seen in small cities and they are characteristic of large cities, because of the critical importance of high land prices for the construction of tall buildings. Usually only office, commercial and hotel users can afford the rents in the city center and thus most tenants of tall buildings are of these classes. Some tall buildings have been built in areas where the bedrock is near surface, because this makes constructing the foundation cheaper.

Today, tall buildings are an increasingly common sight where land is expensive, as in the centers of big cities, because they provide such a high ratio of rentable floor space per unit area of land. One problem with tall buildings is car parking. In the largest cities most people commute via public transport, but for smaller cities a lot of parking spaces are needed. Multi-storey car parks are impractical to build very tall, so a lot of land area is needed. There may be a correlation between a tall building construction and great income inequality but this has not been conclusively proved.

Environmental Impact

The amount of steel, concrete and glass needed to construct a single a tall building is large, and these materials represent a great deal of embodied energy. Tall buildings are thus energy intensive buildings, but tall buildings have a long lifespan. For example the Empire State Building in New York City, the United States completed in 1931 and is still in active use.

Tall buildings have considerable mass, which means that they must be built on a sturdier foundation than would be required for shorter, lighter buildings. Building materials must also be lifted to the top of a tall building during construction, requiring more energy than would be necessary at lower heights. Furthermore, a tall building consumes a lot of electricity because potable and non-potable water have to be pumped

to the highest occupied floors, tall buildings are usually designed to be mechanically ventilated, elevators are generally used instead of stairs, and natural lighting cannot be utilized in rooms far from the windows and the windowless spaces such as elevators, bathrooms and stairwells.

Tall buildings can be artificially lighted and the energy requirements can be covered by renewable energy or other electricity generation of low greenhouse gas emissions. Heating and cooling of tall buildings can be efficient, because of centralized HVAC systems, heat radiation blocking windows and small surface area of the building.

In the lower levels of a tall building a larger percentage of the building cross section must be devoted to the building structure and services than is required for lower buildings: The elevator conundrum creates the need for more lift shafts-everyone comes in at the bottom and they all have to pass through the lower part of the building to get to the upper levels. Building services-power and water enter the building from below and have to pass through the lower levels to get to the upper levels.

In low-rise structures, the support rooms (chillers, transformers, boilers, pumps and air handling units) can be put in basements or roof space-areas which have low rental value. There is, however, a limit to how far this plant can be located from the area it serves. The farther away it is the larger the risers for ducts and pipes from this plant to the floors they serve and the more floor area these risers take. In practice this means that in high-rise buildings this plant is located on plant levels at intervals up the building.

Section B Text Exploration

New Words and Expressions

antennae [æn'teniː]	n.	天线
bundle ['bʌndl]	v.	包，捆，扎，束；把……打成一包（或一捆、一束、一扎）
cantilever ['kæntiliːvə; 'kæntilevə]	n.	悬壁，伸臂
	v.	利用悬臂支撑
chiller ['tʃilə]	n.	冷冻器，冷却器，激冷器
construction [kən'strʌkʃən]	n.	建筑，建造，建设，构造，设计，架设
conundrum [kə'nʌndrəm]	n.	谜语，双关谜，费解的难题，猜不透的难题
conventional [kən'venʃənəl]	a.	习惯的，传统的
dissenting [di'sentiŋ]	a.	持异议的，不同意的，持不同政见的
eliminate [i'limineit]	v.	除掉，除去，消除，排除，根除，摆脱，消灭
embody [im'bɔdi]	v.	使具体化；具体表现，体现，表达
foundation [faun'deiʃən]	n.	建立，设立，创立，创建，创办；（建筑物天然的或人工的）基础，根基，地基；（思想、学说等的）基础，基本原则，根据
girder ['gəːdə]	n.	（大）梁，主梁

Unit 8　Tall Buildings

lateral ['lætərəl]	a.	侧面的，在侧面的，向侧面的，旁边的
intensive [in'tensiv]	a.	加强的；集中的，密集的；深入细致的，透彻的
malleability [ˌmæliə'biliti]	n.	顺从；可锻性，展延性
rationale [ˌræʃə'nɑːli；ˌræʃə'næli]	n.	基本原理，理论基础，理论（或原理）的阐述
renewable [ri'njuːəbl]	a.	可更新的，可继续的，可续借的，可续订的，可再生的
reveted ['rivitid]	a.	铆接的，用铆钉钉牢的；结了婚的
spine [spain]	n.	脊柱，脊椎，脊骨
stairwell ['stɛəwel]	n.	楼梯井
sturdy ['stəːdi]	a.	强壮的，结实的，坚实的，坚固的
suspend [sə'spend]	v.	吊，悬，挂
tubular ['tjuːbjulə]	a.	管的，管状的
vanity ['væniti]	n.	自负，自大；虚荣心，虚夸；无用，无价值；无益的行为，无价值的事物
ventilate ['ventileit]	v.	使通风，使换气
X-bracing	n.	剪刀撑系统；交叉联接；交叉支条

at intervals	每隔一定的时间（或距离），不时，到处
be categorized as	被归类为……
be devoted to	专心于/致力于……的，献身……的
braced frame	撑系框架，刚性构架
cast iron	铸铁
concrete core	混凝土芯，土芯
cut down	缩减，削减
dead load	固定负荷（或负载），持续负载，恒载，底载，静载
elevator shaft	电梯竖井，升降机井

HVAC (high voltage alternating current)	高压交流电
impose on	利用，欺骗，施加影响于
live load	动荷，活负载，有效负载
mass dampers	质量阻尼器
serve as	担任……，充当……，起……的作用
shear wall	剪力墙，抗震墙，耐震壁，风力墙
sky lobby	空中大厅
take on	披上，穿上，戴上；呈现（面貌等），具有（某种性质、特征等）
transfer into	转移，迁移
wind shear	风切变，乱流
wind load	风力载荷

I True and false.

1. A skyscraper tall building is a multi-story building whose architectural height is at most 100 metres or 330 feet. (□T □F)
2. "Vanity height" is the distance between the lowest floor and its architectural top. (□T □F)
3. The tubular systems are fundamental to tall building design. (□T □F)
4. X-bracing, or the "trussed tube", first employed for the John Hancock Center, reduced the lateral load on the building by transferring the load into the interior columns. (□T □F)
5. Tall buildings have considerable mass, which means that they must be built on a sturdier foundation than would be required for shorter, lighter buildings. (□T □F)

II Choose the best answer according to the text.

1. A high-rise is often defined as _____ .

Unit 8 Tall Buildings

  A. any vertical construction for which wind is a more significant load factor than earthquake or weight

  B. any vertical construction for which earthquake or weight is a more significant load factor than wind

  C. a vertical construction for which height is a more important load factor than earthquake or weight

  D. a vertical construction for which height is the most important load factor

2. The design and construction of tall buildings involves the following aspects except _____.

  A. creating safe, habitable spaces

  B. supporting their weight

  C. resisting wind and earthquakes

  D. protecting occupants from rain

3. Vertical supports can come in several types, among which the most common for tall buildings can NOT be categorized as _____.

  A. steel frames   B. concrete cores

  C. window design   D. shear walls

4. A steel frame eliminated the inefficient part of a shear wall and consolidated support members in a much stronger fashion by _____.

  A. allowing the horizontal supports throughout

  B. allowing both horizontal and vertical supports throughout

  C. allowing the vertical supports throughout

  D. allowing the symmetrical supports throughout

5. _____ is fundamental to tall building design.

  A. The elevator conundrum

  B. Loading and vibration

  C. Steel frame

  D. The tubular system

III Translation.

1. 高层建筑的一个共同特点就是幕墙悬空于一个钢架结构，而不是传统建筑的承重墙。

  A. One common character of tall buildings is a steel framework from which load-bearing walls of conventional construction are suspended, rather than curtain walls.

B. One common trait of tall buildings is an iron framework from which curtain walls are suspended, rather than load-bearing walls of conventional construction.

C. One common property of tall buildings is an iron framework from which load-bearing walls of conventional construction are suspended, rather than curtain walls.

D. One common feature of tall buildings is a steel framework from which curtain walls are suspended, rather than load-bearing walls of conventional construction.

2. 就技术而言，静荷载，即结构负荷，大于活荷载，即结构中物品的重量。

A. From the perspective of technology, the live load, the load of the structure, is larger than the dead load, the weight of things in the structure.

B. As for the technology, the inactive load, the load of the structure, is larger than the active load, the weight of things in the structure.

C. In technical terms, the dead load, the load of the structure, is larger than the live load, the weight of things in the structure.

D. Concerning technology, the dead load, the load of the structure, is smaller than the live load, the weight of things in the structure.

3. 鉴于大多数人不会一次攀爬几层楼梯，电梯的发明是建造高层建筑的先决条件。

A. Because most people would not climb no more than a few flights of stairs at a time, the invention of the elevator was a precondition for the invention of tall buildings.

B. The invention of the elevator was a precondition for the invention of tall buildings, given that most people would not climb more than a few flights of stairs at a time.

C. The invention of the elevator was a presupposition for the invention of tall buildings, considering that most people would climb more than a few flights of stairs at a time.

D. For few people would not climb more than a few flights of stairs at a time, the invention of the elevator was a precondition for the invention of tall buildings.

4. 高层建筑可采用人工照明，其能源需求可以由可再生能源或其他温室气体排放量低的发电方式满足。

A. Tall buildings can be naturally lighted and the energy needs can be satisfied by recycled energy or other electricity generation of low greenhouse gas production.

B. Tall buildings can be naturally lighted and the recycled energy or other electricity generation of low greenhouse gas release can satisfy the energy demand.
C. Tall buildings can be artificially lighted and the renewable energy or other electricity generation of low greenhouse gas emissions can replace the energy demand.
D. Tall buildings can be artificially lighted and the energy requirements can be covered by renewable energy or other electricity generation of low greenhouse gas emissions.

Section C Supplementary Reading

Burj Khalifa

The Burj Khalifa is a megatall skyscraper in Dubai, the United Arab Emirates. With a total height of 829.8 m (2,722 ft) including the antenna and a roof height of 828 m (2,717 ft), the Burj Khalifa is currently the tallest structure in the world since topping out in late 2008.

Construction of the Burj Khalifa began in 2004, with the exterior completed 5 years later in 2009. The primary structure is reinforced concrete. The building was opened in 2010 as part of a new development called Downtown Dubai. It is designed to be the centrepiece of large-scale, mixed-use development. The decision to construct the building is reportedly based on the government's decision to diversify from an oil-based economy, and for Dubai to gain international recognition. The building was named in honour of the ruler of Abu Dhabi and president of the United Arab Emirates, Khalifa bin Zayed Al Nahyan; Abu Dhabi and the UAE government lent Dubai money to pay its debts. The building broke numerous height records, including its designation as the tallest tower in the world. Burj Khalifa was designed by Adrian Smith, then of Skidmore, Owings & Merrill (SOM), whose firm designed the Willis Tower and One World Trade Center. Hyder Consulting was chosen to be the supervising engineer with NORR Group Consultants International Limited chosen to supervise the architecture of the project. The design is derived from the Islamic architecture of the region, such as in the Great Mosque of Samarra. The Y-shaped tripartite floor geometry is designed to optimize residential and  hotel space. A buttressed central core and wings are used to support the height of the building. Although this design was derived from Tower Palace III, the Burj Khalifa's central core houses all vertical transportation with the exception of egress stairs within each of the wings. The structure also features a cladding system which is designed to withstand Dubai's hot summer temperatures. It contains a total of 57 elevators and 8 escalators. Critical reception to Burj Khalifa has been generally positive, and the building has received many awards. However, the labour issues during construction were controversial, since the building was built primarily by migrant workers from South Asia with several allegations of mistreatment. Poor working conditions are common, as the result of the lack of minimum wage laws in the United Arab Emirates. Several instances of suicides have been reported, which is not uncommon for migrant construction workers in Dubai despite safety precautions in place.

Conception

Burj Khalifa was designed to be the centrepiece of a large-scale, mixed-use development that would include 30,000 homes, nine hotels (including The Address Downtown Dubai), 3 hectares (7.4 acres) of parkland, at least 19 residential towers, the Dubai Mall, and the 12-hectare (30-acre) artificial Burj Khalifa Lake. The decision to build Burj Khalifa is reportedly based on the government's decision to diversify from an oil-based economy to one that is service and tourism based. According to officials, it is necessary for projects like Burj Khalifa to be built in the city to garner more international recognition, and hence investment. "He (Sheikh Mohammed bin Rashid Al Maktoum) wanted to put Dubai on the map with something really sensational," said Jacqui Josephson, a tourism and VIP delegations executive at Nakheel Properties. The tower was known as Burj Dubai ("Dubai Tower") until its official opening in January 2010. It was renamed in honour of the ruler of Abu Dhabi and president of the United Arab Emirates,

Khalifa bin Zayed Al Nahyan; Abu Dhabi and the federal government of UAE lent Dubai tens of billions of USD so that Dubai could pay its debts-Dubai borrowed at least $80 billion for construction projects. In the 2000s, Dubai started diversifying its economy but it suffered from an economic crisis in 2007-2010, leaving large-scale projects already in construction abandoned.

The tower was designed by Skidmore, Owings and Merrill (SOM), who also designed the Willis Tower (formerly the Sears Tower) in Chicago and the One World Trade Center in New York City. Burj Khalifa uses the bundled tube design of the Willis Tower, invented by FazlurRahman Khan. Proportionally, the design uses half the amount of steel used in the construction of the Empire State Building thanks to the tubular system. Its design is reminiscent of Frank Lloyd Wright's vision for The Illinois, a mile-high skyscraper designed for Chicago. According to Marshall Strabala, a SOM architect who worked on the building's design team, Burj Khalifa was designed based on the 73rd floor Tower Palace Three, an all residential building in Seoul. In its early planning, Burj Khalifa was intended to be entirely residential.

Subsequent to the original design by Skidmore, Owings and Merrill, Emaar Properties chose Hyder Consulting to be the supervising engineer with NORR Group Consultants International Ltd chosen to supervise the architecture of the project. Hyder was selected for their expertise in structural and MEP (mechanical, electrical and plumbing) engineering. Hyder Consulting's role was to supervise construction, certify SOM's design, and be the engineer and architect of record to the UAE authorities. NORR's role was the supervision of all architectural components including on site supervision during construction and design of a 6-storey addition to the Office Annex Building for architectural documentation. NORR was also responsible for the architectural integration drawings for the Armani Hotel included in the Tower. Emaar Properties also engaged GHD, an international multidisciplinary consulting firm, to act as an independent verification and testing authority for concrete and steelwork.

The design is derived from Islamic architecture. As the tower rises from the flat desert base, there are 27 setbacks in a spiralling pattern, decreasing the cross section of the tower as it reaches toward the sky and creating convenient outdoor terraces. These setbacks are arranged and aligned in a way that minimizes vibration wind loading from eddy currents and vortices. At the top, the central core emerges and is sculpted to form a finishing spire. At its tallest point, the tower sways a total of 1.5 m (4.9 ft).

As part of a study which reveals the unnecessary "vanity space" added to the top of the world's tallest buildings by the Council on Tall Buildings and Urban Habitat (CTBUH), it was revealed that without its 244-metre spire, the 828-metre Burj Khalifa would drop to a substantially smaller 585-metre height without any reduction in usable space. As the report states, the spire "could be a skyscraper on its own".

The spire of Burj Khalifa is composed of more than 4,000 tonnes (4,400 short tons; 3,900 long tons) of structural steel. The central pinnacle pipe weighs 350 tonnes (390 short tons; 340 long tons) and has a height of 200 m (660 ft). The spire also houses communications equipment. In 2009, architects announced that more than 1,000 pieces of art would adorn the interiors of Burj Khalifa, while the residential lobby of Burj Khalifa would display the work of JaumePlensa.

The cladding system consists of 142,000 m² (1,528,000 sq ft) of more than 26,000 reflective glass panels and aluminium and textured stainless steel spandrel panels with vertical tubular fins. The architectural glass provides solar and thermal performance as well as an anti-glare shield for the intense desert sun, extreme desert temperatures and strong winds. In total the glass covers more than 174,000 m² (1,870,000 sq ft). The exterior temperature at the top of the building is thought to be 6 °C (11 °F) cooler than at its base.

A 304-room Armani Hotel, the first of four by Armani, occupies 15 of the lower 39 floors. The hotel was supposed to open on 18 March 2010, but after several delays, it finally opened to the public on 27 April, 2010. The corporate suites and offices were also supposed to open from March onwards, yet the hotel and observation deck remained the only parts of the building which were open in April 2010.

The sky lobbies on the 43rd and 76th floors house swimming pools. Floors through to 108 have 900 private residential apartments (which, according to the developer, sold out within eight hours of being on the market). An outdoor zero-entry swimming pool is located on the 76th floor of the tower. Corporate offices and suites fill most of the remaining floors, except for a 122nd, 123rd and 124th floor where the atmosphere restaurant, sky lobby and an indoor and outdoor observation deck is located respectively.

A total of 57 elevators and 8 escalators are installed. The elevators have a capacity of 12 to 14 people per cabin, the fastest rising and descending at up to 10 m/s (33 ft/s) for double-deck elevators. However, the world's fastest single-deck elevator still belongs to Taipei 101 at 16.83 m/s (55.2 ft/s). Engineers had considered installing the world's first triple-deck elevators, but the final design calls for double-deck elevators. The double-deck elevators are equipped with entertainment features such as LCD displays to serve visitors during their travel to the observation deck. The building has 2,909 stairs from the ground floor to the 160th floor.

The graphic design identity work for Burj Khalifa is the responsibility of Brash Brands, an independent international creative branding agency based in London. Design of the global launch events, communications, and visitors' centres for Burj Khalifa have also been created by Brash Brands as well as the roadshow exhibition for the Armani Residences, which are part of the Armani Hotel within Burj Khalifa, which toured Milan, London, Jeddah, Moscow and Delhi.

Plumbing Systems

The Burj Khalifa's water system supplies an average of 946,000 L (250,000 US gal) of water per day through 100 km (62 mi) of pipes. An additional 213 km (132 mi) of piping serves the fire emergency system, and 34 km (21 mi) supplies chilled water for the air conditioning system. The waste water system uses gravity to discharge water from plumbing fixtures, floor drains, mechanical equipment and storm water, to the city municipal sewer.

Air Conditioning

The air conditioning has been provided by Voltas. The air conditioning system draws air from the upper floors where the air is cooler and cleaner than on the ground. At peak cooling times, the tower's cooling is equivalent to that provided by 13,000 short tons (26,000,000 lb) of melting ice in one day, or about 46 MW. Water is collected via a condensate collection system and is used to irrigate the nearby park.

Window Cleaning

To wash the 24,348 windows, totaling 120,000 m^2 (1,290,000 sq ft) of glass, the building has three horizontal tracks which each hold a 1,500 kg (3,300 lb) bucket machine. Above level 109, and up to tier 27, traditional cradles from davits are used. The top of the building is cleaned by a crew who use ropes to descend from the top to gain access. Under normal conditions, when all building maintenance units are operational, it takes 36 workers three to four months to clean the entire exterior façade.

Unmanned machines will clean the top 27 additional tiers and the glass spire. The cleaning system was developed in Melbourne, Australia at a cost of 8 million Australian dollars. The contract for building the state-of-the-art machines was won by Australian company CoxGomyl, a renowned manufacturer of Building Maintenance Units.

Unit 9
Bridge Works

Section A Text

Bridges

A bridge is a structure built to span physical obstacles without closing the way underneath such as a body of water, valley, or road, for the purpose of providing passage over the obstacle. Designs of bridges vary depending on the function of the bridge, the nature of the terrain where the bridge is constructed and anchored, the material used to make it, and the funds available to build it.

The first bridges made by humans were probably spans of cut wooden logs or planks and eventually stones, using a simple support and crossbeam arrangement. Some early Americans used trees or bamboo poles to cross small caverns or wells to get from one place to another. A common form of lashing sticks, logs, and deciduous branches together involved the use of long reeds or other harvested fibers woven together to form a huge rope capable of binding and holding together the materials used in early bridges.

Bridges can be categorized in several different ways. Common categories include the type of structural elements used, by what they carry, whether they are fixed or movable, and by the materials used.

Structure Type

Bridges may be classified by how the forces of tension, compression, bending, torsion and shear are distributed through their structure. Most bridges will employ all of the principal forces to some degree, but only a few will predominate. The separation of forces may be quite clear. In a suspension or cable-stayed span, the elements in tension are distinct in shape and placement. In other cases the forces may be distributed among a large number of members, as in a truss.

Beam bridges are horizontal beams supported at each end by substructure units and can be either simply supported when the beams only connect across a single span, or continuous when the beams are connected across two or more spans. When there are multiple spans, the intermediate supports are known as piers. The earliest beam bridges were simple logs that sat across streams and similar simple structures. In modern times, beam bridges can range from small, wooden beams to large, steel boxes. The vertical force on the bridge becomes a shear and flexural load on the beam which is transferred down its length to the substructures on either side. They are typically made of steel, concrete or wood. Beam bridge spans rarely exceed 250 ft (76 m) long, as the flexural stresses increase proportional to the square of the length (and deflection increases proportional to the 4th power of the length). However, the main span of the Rio-Niteroi Bridge, a box girder bridge, is 980 ft (300 m).

A truss bridge is a bridge whose load-bearing superstructure is composed of a truss. This truss is a structure of connected elements forming triangular units. The connected elements (typically straight) may be stressed from tension, compression, or sometimes both in response to dynamic loads. Truss bridges are one of the oldest types of modern bridges. The basic types of truss bridges shown in this article have simple designs which could be easily analyzed by nineteenth and early twentieth century engineers. A truss bridge is economical to construct

owing to its efficient use of materials.

Cantilever bridges are built using cantilevers—horizontal beams supported on only one end. Most cantilever bridges use a pair of continuous spans that extend from opposite sides of the supporting piers to meet at the center of the obstacle the bridge crosses. Cantilever bridges are constructed using much the same materials and techniques as beam bridges. The difference comes in the action of the forces through the bridge.

Some cantilever bridges also have a smaller beam connecting the two cantilevers, for extra strength. Arch bridges have abutments at each end. The weight of the bridge is thrust into the abutments at either side. The earliest known arch bridges were built by the Greeks, and include the Arkadiko Bridge.

With the span of 220 m (720 ft), the Solkan Bridge over the Sòca River at Solkan in Slovenia is the second largest stone bridge in the world and the longest railroad stone bridge. It was completed in 1905. Its arch, which was constructed from over 5,000 tonnes (4,900 long tons; 5,500 short tons) of stone blocks in just 18 days, is the second largest stone arch in the world, surpassed only by the Friedensbrücke in Plauen, and the largest railroad stone arch. The arch of the Friedensbrücke, which was built in the same year, has the span of 90 m (295 ft) and crosses the valley of the Syrabach River. The difference between the two is that the Solkan Bridge was built from stone blocks, whereas the Friedensbrücke was built from a mixture of crushed stone and cement mortar.

Tied arch bridges have an arch-shaped superstructure, but differ from conventional arch bridges. Instead of transferring the weight of the bridge and traffic loads into thrust forces into the abutments, the ends of the arches are restrained by tension in the bottom chord of the structure. They are also called bowstring arches.

Suspension bridges are suspended from cables. The earliest suspension bridges were made of ropes or vines covered with pieces of bamboo. In modern bridges, the cables hang from towers that are attached to caissons or cofferdams. The caissons or cofferdams are implanted deep into the bed of the lake, river or sea. Sub-types include the simple

suspension bridge, the stressed ribbon bridge, the underspanned suspension bridge, the suspended-deck suspension bridge, and the self-anchored suspension bridge. There is also what is sometimes called a "semi-suspension" bridge, of which the Ferry Bridge in Burton-upon-Trent is the only one of its kind in Europe. Cable-stayed bridges, like suspension bridges, are held up by cables. However, in a cable-stayed bridge, less cables are required and the towers holding the cables are proportionately higher.

Fixed or Movable Bridges

Most bridges are fixed bridges, meaning they have no moving parts and stay in one place until they fail or are demolished. Temporary bridges, such as Bailey bridges, are designed to be assembled, and taken apart, transported to a different site, and re-used. They are important in military engineering, and are also used to carry traffic while an old bridge is being rebuilt. Movable bridges are designed to move out of the way of boats or other kinds of traffic, which would otherwise be too tall to fit. These are generally electrically powered.

Double-decked Bridges

Double-decked (or double-decker) bridges have two levels, such as the George Washington Bridge, connecting New York City to Bergen County, New Jersey, USA, as the world's busiest bridge, carrying 102 million vehicles annually; truss work between the roadway levels provided stiffness to the roadways and reduced movement of the upper level when the lower level was installed three decades after the upper level. The Tsing Ma Bridge and Kap Shui Mun Bridge in Hong Kong have six lanes on their upper decks, and on their lower decks there are two lanes and a pair of tracks for MTR metro trains. Some double-decked bridges only use one level for street traffic; the Washington Avenue Bridge in Minneapolis

reserves its lower level for automobile and light rail traffic and its upper level for pedestrian and bicycle traffic. Likewise, in Toronto, the Prince Edward Viaduct has five lanes of motor traffic, bicycle lanes, and sidewalks on its upper deck; and a pair of tracks for the Bloor-Danforth subway line on its lower deck. The western span of the San Francisco-Oakland Bay Bridge also has two levels.

Robert Stephenson's High Level Bridge across the River Tyne in Newcastle, completed in 1849, is an early example of a double-decked bridge. The upper level carries a railway, and the lower level is used for road traffic. Other examples include Britannia Bridge over the Menai Strait and Craigavon Bridge in Derry, Northern Ireland. The Oresund Bridge between Copenhagen and Malmö consists of a four-lane highway on the upper level and a pair of railway tracks at the lower level. Tower Bridge in London is different example of a double-decked bridge, with the central section consisting of a low level bascule span and a high level footbridge.

Three-way Bridges

A three-way bridge has three separate spans which meet near the center of the bridge. The bridge appears as a "T" or "Y" when viewed from above. Three-way bridges are extremely rare. The Tridge, Margaret Bridge, and Zanesville Y-Bridge are examples.

Bridge Types by Use

A bridge can be categorized by what it is designed to carry, such as trains, pedestrian or road traffic, a pipeline or waterway for water transport or barge traffic. An aqueduct is a bridge that carries water, resembling a viaduct, which is a bridge that connects points of equal height. A road-rail bridge carries both road and rail traffic.

Some bridges accommodate other purposes, such as the tower of Novy Most Bridge in Bratislava, which features a restaurant, or a bridge-restaurant which is a bridge built to serve as a restaurant. Other suspension bridge towers carry transmission antennas.

Bridges are subject to unplanned uses as well. The areas underneath some bridges have become makeshift shelters and homes to homeless people, and the undersides of bridges all around the world are spots of prevalent graffiti. Some bridges attract people attempting suicide, and become known as suicide bridges

Bridge Types by Material

The materials used to build the structure are also used to categorize bridges. Until the end of the 18th Century, bridges were made out of timber, stone and masonry. Modern bridges are currently built in concrete, steel, fiber reinforced polymers (FRP), stainless steel or combinations of those materials. Living bridges have been constructed of live plants such as tree roots in India and vines in Japan.

Bridge Type	Materials Used
Cantilever	For small footbridges, the cantilevers may be simple beams; however, large cantilever bridges designed to handle road or rail traffic use trusses built from structural steel, or box girders built from prestressed concrete.
Suspension	The cables are usually made of steel cables galvanized with zinc, along with most of the bridge, but some bridges are still made with steel reinforced concrete.
Arch	Stone, brick and other such materials that are strong in compression and somewhat so in shear.
Beam	Beam bridges can use pre-stressed concrete, an inexpensive building material, which is then embedded with rebar. The resulting bridge can resist both compression and tension forces.
Truss	The triangular pieces of truss bridges are manufactured from straight and steel bars, according to the truss bridge designs.

Unit 9　Bridge Works

Section B　Text Exploration

New Words and Expressions

abutment [ə'bʌtmənt]	n.	拱座，支座，支柱，支墩，桥墩，桥台，扶垛，扶壁
anchor ['æŋkə]	v.	抛锚泊（船），抛锚，锚泊，泊船；使固定，使稳固，把……扎牢（或扣牢、系住、粘住）
bend [bend]	v.	使（弓）弯，弯（弓）；挽，拉（弓）
caisson [kə'su:n; 'keisən]	n.	沉箱，潜水箱
cavern ['kævən]	n.	洞穴，山洞（尤指大洞穴、大山洞）
cofferdam ['kɔfə‚dæm]	n.	（河流、湖泊等的）围堰；潜水箱，沉箱
deciduous [di'sidjuəs]	a.	（树和灌木）每年落叶的，（森林等）有落叶树的
deflection [di'flekʃən]	n.	转向，偏斜，歪斜，挠曲
demolish [di'mɔliʃ]	v.	拆除，拆毁（建筑物或其他结构），爆破
flexural ['flekʃərəl]	a.	弯曲的，曲折的
galvanize ['gælvənaiz]	v.	通电流于，给……通电
graffiti [grə'fi:ti]	n.	墙上乱写乱画的东西（单数为 graffito）
lashing ['læʃiŋ]	n.	（用绳子等）捆绑，捆扎；（捆绑用的）绳子，绳索
rebar ['ri:ba:]	n.	钢筋，螺纹钢（筋）
reed [ri:d]	n.	苇子，芦秆（或茅草等）

shear [ʃiə]		v.	修剪掉，剪掉，割去
		n.	剪力，剪应力；切变，剪应变
pier [piə]		n.	桥墩，墩；（凸式）码头，直码头；防波堤，突堤；脚柱，支柱，垂直的支承结构构件，结构底座
span [spæn]		n.	跨度；一段时间
suicide ['sjuisaid]		n.	自杀，自毁，自灭
terrain [te'rein; 'terein]		n.	地面，地带，地域，地区，地貌
triangular [trai'æŋgjulə]		a.	三角（形）的，三个角的；三者间的，三人间的，三方，三国间的
viaduct ['vaiədʌkt]		n.	高架桥，跨线桥；高架铁路，高架公路

be attached to	附属于，系于；爱慕
be composed of	由……组成
be embedded with	镶有……，嵌有……
be surpassed by	被……超越
be thrust into	被推入……
be woven together	被交织在一起
bowstring arch bridge	系杆拱桥，弓弦拱桥
box girder bridge	箱形梁桥，箱梁桥
cable-stayed bridge	索拉桥
cement mortar	水泥砂浆，水泥灰浆
double-decked bridge	双层桥
fiber reinforced polymers (FRP)	纤维增强聚合物
in compression	受压
MTR (Mass Transit Railway)	港铁
prestressed concrete	预应力混凝土
suspended deck	悬桥面
suspension bridge	吊桥
tied arch bridge	系杆拱桥

Unit 9 Bridge Works

Exercises

I True and false.

1. In modern times, beam bridges can range from small, wooden beams to large, steel boxes. (□T □F)
2. A truss bridge is a bridge whose load-bearing superstructure is made up of a truss. (□T □F)
3. Cantilever bridges are built using cantilevers—horizontal beams supported on two ends. (□T □F)
4. The stressed ribbon bridge does not belong to suspension bridge. (□T □F)
5. A three-way bridge has three separate spans which meet near the end of the bridge. (□T □F)

II Choose the best answer according to the text.

1. _____ have no moving parts and stay in one place until they fail or are demolished.
   A. Suspension bridges   B. Fixed bridges
   C. Arch bridges   D. Cantilever bridges
2. Modern bridges are currently built in the following materials except _____.
   A. concrete   B. mud
   C. fiber reinforced polymers (FRP)   D. stainless steel
3. Which expression is true according to the text?
   A. Large cantilever bridges designed to handle road or rail traffic use trusses built from structural steel, or box girders built from prestressed concrete.
   B. The cables are usually made of steel cables galvanized with copper.
   C. Beam bridges can use pre-stressed concrete, an inexpensive building material, which is then embedded with rebar.
   D. The rectangular pieces of truss bridges are manufactured from straight and steel bars.
4. Arkadiko Bridge is a kind of _____.

A. stone bridge B. truss bridge
C. suspension bridge D. arch bridge

5. Robert Stephenson's High Level Bridge is an example of _____ .
A. three-way bridge B. fixed bridge
C. double-decked bridge D. movable bridge

III Translation.

1. 桥是一种跨越物理障碍物的建筑物，不阻碍下面的河流、峡谷或道路，而是为跨越这些障碍提供通道。
   A. A bridge is a structure built to span physical obstacles by closing the way underneath such as a body of water, valley, or road, for the purpose of providing passage over the obstruction.
   B. A bridge is a long road built to span physical obstacles without closing the way underneath such as a body of water, valley, or road, for the purpose of providing passage over the obstruction.
   C. A bridge is a long road built to span physical obstacles without closing the way above such as a body of water, valley, or road, for the purpose of providing passage over the obstacle.
   D. A bridge is a structure built to span physical obstacles without closing the way underneath such as a body of water, valley, or road, for the purpose of providing passage over the obstacle.

2. 桥梁的设计会根据桥梁的功能、桥梁建筑和固定区域的地形特质、桥梁的建筑材料及可用资金而变化。
   A. Designs of bridges change according to the height of the bridge, the nature of the terrain where the bridge is constructed and fixed, the material used to make it, and the funds available to build it.
   B. Designs of bridges vary based on the function of the bridge, road information where the bridge is constructed and anchored, the material used to make it, and the funds available to build it.
   C. Designs of bridges vary depending on the function of the bridge, the nature of the terrain where the bridge is constructed and anchored, the material used to make it, and the funds available to build it.
   D. Designs of bridges vary according to the function of the bridge, the nature of the

terrain where the bridge is built and anchored, the material used to make it, and the funds reserved to build it.

3. 高架渠是一种可以输送水的桥梁，类似于连接相同高度点的高架桥。
   A. An aqueduct is a bridge that carries water, resembling a viaduct, which is a bridge that joints points of various heights.
   B. An aqueduct is a bridge that carries water, resembling a viaduct, which is a bridge that connects points of equal height.
   C. A viaduct is a bridge that carries water, like an aqueduct, which is a bridge that connects points of various heights.
   D. A viaduct is a bridge that carries water, like an aqueduct, which is a bridge that joints points of equal height.

4. 桥梁上的垂直的力成为梁柱上的一种剪切和弯曲的负载，这种力可以向下转移到两侧的子结构。
   A. The horizontal force on the bridge becomes a shear and flexible load on the beam that is transferred down its width to the substructures on either side.
   B. The horizontal force on the bridge becomes a shear and flexural load on the beam, which is transferred down its length to the structures on neither side.
   C. The vertical force on the bridge becomes a shear and flexural load on the beam which is transferred down its length to the substructures on either side.
   D. The vertical force on the bridge becomes a shear and flexible load on the beam, which is transferred down its width to the structures on neither side.

Section C Supplementary Reading

Bridge Design

Data Needed for Designing a Bridge

A plan of the site showing all obstacles to be bridged such as rivers, streets, roads or railroads, the contour lines of valleys and the desired alignment of the new traffic route.

Longitudinal section of the ground along the axis of the planned bridge with the conditions for clearances or required flood widths.

Desired vertical alignment of the new route.

Required width of the bridge, width of lanes, median, walkways, safety rails etc.

Soil conditions for foundations, results of borings with a report on the geological situation and soil mechanics data. The degree of difficulty of foundation work has a considerable influence on the choice of the structural system and on the economical span length.

Local conditions like accessibility for the transport of equipment, materials and structural elements. Which materials are available and economical in that part of the country? Is water or electric power at hand? Can a high standard of technology be used or must the bridge be built with primitive methods and a small number of skilled laborers?

Weather and environmental conditions, floods, high and low tide levels, periods of drought, range of temperatures, and length of frost periods.

Topography of the environment—open land, flat or mountainous land, scenic country. Town with small old houses or city with high-rise buildings. The scale of the environment has an influence on the design.

Environmental requirements regarding aesthetic quality. Bridges in towns that affect the urban environment and that are frequently seen at close range—especially pedestrian bridges—need more delicate shaping and treatment than bridges in open country. Is protection of pedestrians against spray and noise needed? Is noise protection necessary for houses close to the bridge?

Rough Sketch of Bridge Design

When the engineer is sure that a design idea has emerged in his mind, he should pick up a pencil and a scale and by the help of sketching, learned at school, he should start from sketching the probable road direction, beam depth (for beam bridge), the piers, the abutments and the bottom edge of the beam is

drawn.

For a heavily funded project, high slenderness ratio is preferable otherwise if the decisive factor is the cost then slenderness ratio can be reduced. Analyze the sketch critically for the proportion between the spans, the clearance under the beam, soil conditions around the piers and the abutments, the adaptiveness of the piers to the surroundings, no. of piers and for the curvature of the vertical alignment. More than one sketch may follow after the critical analysis with changes in the design and supporting conditions. Work out the alternatives, discuss with colleagues, architects advisors and the client to draw out a final sketch.

Only now should calculations begin, and in the first place with simple and rough approximations to check whether the assumed dimensions will be sufficient and whether the necessary sectional areas of reinforcing steel or of pre-stressing tendons will leave sufficient space, to allow the concrete to be placed and compacted without difficulty. Then some runs with computer programs can be made, using different depths or other variables in order to find the most economical dimensions; these should, however, only be chosen if no other essential requirements, such as aesthetics, length of approaches, grades etc. are affected.

Once the designer or the design team has made its choice, then the principle design drawings with all dimensions and explanations can be drawn up for approval of the authorities. As the map alone is not sufficient to clearly show the locality and impact on the environment and appearance so a model or some well shot photos can help the citizens, client and critics to realize the existence of bridge.

Finalizing the Bridge Design

After the approval of the design, the final design work can begin with rigorous calculations of forces, stresses etc. for all kinds of loads or attacks and then the structural detailing has to be done. The scaffolding and equipment, which will be needed for the construction of the particular type of bridge, also has to be worked out. Numerous drawings and tables with thousands of numbers and figures for all dimensions, sizes and levels must be made with specifications for

the required type and quality of the building materials. This phase entails the greatest amount of work for the bridge engineer, and calls for considerable knowledge and skill.

The Mighty Mac: A sublime Engineering Feat

The "Mighty Mac", the strong, graceful, bridge, which spans the Straits of Mackinac in Michigan, is one of the nation's greatest bridge-building achievements of the 20th century, uniquely conquering wind, waves and ice over its 5-mi (8 km) span. The length of the entire Mighty Mac, including 30 truss span bridges joining the suspension bridge to land, is 26,242.7 ft (7,999 m).

Designed to withstand an "infinite" amount of wind, plus other forces of nature, the Mighty Mac also has vastly improved commerce between Michigan's Upper Peninsula and Lower Peninsula, which had previously been almost two separate states economically, politically and culturally. Linking the two peninsulas together has been by far the greatest accomplishment of the Mackinac Bridge. It really opened up the Upper Peninsula to much greater tourism. The logging industry has benefited greatly from it as well. Dubbed the Mighty Mac since its early days, the bridge also has drawn many more visitors to the Straits area, which became a premiere resort destination for the rich in the 1880s.

Travel between the Upper Peninsula and Lower Peninsula was difficult, sometimes impossible, without a bridge. The Straits were often locked in ice during the winter, so one walked, rode by horse, or took a sleigh or dog team. Some travelers froze to death trying.

The state of Michigan first tried to handle the increasing demand with eight ferries. Sometimes as many as 800 waiting cars were lined up for 5 mi. As the delays got worse, people got serious about a bridge. Completion

of the Golden Gate Bridge in San Francisco in 1937 gave planners greater incentive.

The obstacles, however, were intimidating. In 1820, Henry Schoolcraft, a member of a scientific expedition, which the War Department sent to study the Straits area, noted cavities in the rock. Some geologists in the 1930s said this rock couldn't support the weight of a bridge.

The north-south bridge also would have to stand broadside to winter gales, which swept in from the Great Lakes as if in a wind tunnel. (Wind speeds actually reached 124 mi per hour in May 2003.) Not least, the bridge would have to withstand swirling currents and tremendous pressure from ice, which could be eight or more feet thick in the shipping channels. The entire bridge, including truss spans, also would have to be 5 mi (8 km) long. Quite a few engineers said it couldn't be built.

Then there was the cost. The first bridge authority, created in 1934, made a bid for depression-era federal financing. It estimated a total cost of $35 million. The government rejected the bid. Former governor Chase Osborn, who had once opposed the bridge idea, vehemently supported it by 1935.

"Suppose on a trunk line—and the Straits road is that—there was a mud hole or chasm or abyss or sink-hole eight miles wide that every car had to be pulled over or across or through," he wrote that year. "Something would be done about that at once."

Osborn articulated a vision of the economic development, which would result from joining the Upper Peninsula with the northern part of the Lower Peninsula. "Michigan is unifying itself," he wrote, "and a magnificent new route through Michigan to Lake Superior and the Northwest United States is developing, via the Straits of Mackinac. It cannot continue to grow as it ought with clumsy and inadequate ferries for any portion of the year."

Many also recognized that a bridge would spur tourism, Michigan's second most important industry. W. Stewart Woodfill, manager of the Grand Hotel, closed the hotel at the end of the 1949 season and lobbied for the bridge during the winter, 1950, session of the state legislature in Lansing. It paid off. In 1950, the legislature created a new publicly owned Mackinac Bridge Authority and appropriated funds for preliminary studies and surveys.

Results showed it wouldn't be cheap. On Dec. 17, 1953, bids were accepted from a group of underwriters for the sale of $99,800,000 in bridge authority bonds, which investors throughout the United States subsequently purchased.

Designed by the noted Engineer David B. Steinman, Its 8,614-ft-long suspension span between anchorages surpasses The Golden Gate Bridge's 6,450-ft-long (1,966 m) suspension span, though the Golden Gate's central span is longer — 4,200 ft (1,280.2 m) compared with Mackinac's 3,800 ft (1,158.2 m).

"The way the bridge is built, with a North-South orientation, means the wind comes straight through the Straits and hits it sideways," Sweeney said. "The wind speed, current in the Straits, and the massive ice, which builds up were three of the most challenging features in building the bridge. The bridge can withstand an infinite amount of wind in the summer. During the winter, we assume that several features of the bridge are plugged with ice and snow, reducing the wind-failure speed down to just more than 600 miles per hour."

"Ice in the channel, and just outside the channel, builds up to about eight feet thick. That was one of the reasons why massive caissons around the two main tower piers are approximately 10 feet below the surface, allowing the ice to move back and forth without putting pressure on the caissons."

To produce a titanic bridge withstanding these forces, Steinman designed special features in suspension bridge aerodynamics. The bridge is absolutely windproof in terms of stability against all types of oscillations.

For strength, the bridge protects and strengthens support piers with caissons and cofferdams. Extensive geological studies, and load tests on the rock under water at the construction site had shown that the rock was not weak because of cavities. Results showed, in fact, that even the weakest rock could support more than 60 tons per square, more than four times the greatest possible load from the bridge structure.

Unit 10
Road Works

Section A Text

Highway Engineering

Highway engineering is an engineering discipline branching from civil engineering that involves the planning, design, construction, operation, and maintenance of roads, bridges, and tunnels to ensure safe and effective transportation of people and goods. Highway engineering became prominent towards the latter half of the 20th century after World War II. Standards of highway engineering are continuously being improved. Highway engineers must take into account future traffic flows, design of highway intersections, geometric alignment and design, highway pavement materials and design, structural design of pavement thickness, and pavement maintenance.

The beginning of road construction could be dated to the time of the Romans. With the advancement of technology from carriages pulled by two horses to vehicles with power equivalent to 100 horses, road development had to follow suit. The construction of modern highways did not begin until the late 19th to early 20th century. The first research dedicated to highway engineering was initiated in the United Kingdom with the introduction of the Transport Research Laboratory (TRL) in 1930. In the USA, highway engineering became an important discipline with the passing of the Federal-Aid Highway Act of 1944, which aimed to connect 90% of cities with a population of 50,000 or more. With constant stress from vehicles which

grew larger as time passed, improvements to pavements were needed. With technology out of date, in 1958 the construction of the first motorway in Great Britain (the Preston bypass) played a major role in the development of new pavement technology.

Developed countries are constantly faced with high maintenance cost of aging transportation highways. The growth of the motor vehicle industry and accompanying economic growth has generated a demand for safer, better performing, less congested highways. The growth of commerce, educational institutions, housing, and defense have largely drawn from government budgets in the past, making the financing of public highways a challenge. The multipurpose characteristics of highways, economic environment, and the advances in highway pricing technology are constantly changing. Therefore, the approaches to highway financing, management, and maintenance are constantly changing as well.

The economic growth of a community is dependent upon highway development to enhance mobility. However, improperly planned, designed, constructed, and maintained highways can disrupt the social and economic characteristics of any size community. Common adverse impacts to highway development include damage of habitat and biodiversity, creation of air and water pollution, noise/vibration generation, damage of natural landscape, and the destruction of a community's social and cultural structure. Highway infrastructure must be constructed and maintained to high qualities and standards.

There are three key steps for integrating environmental considerations into the planning, scheduling, construction, and maintenance of highways. This process is known as an Environmental Impact Assessment, or EIA, as it systematically deals with the following elements: (1) identification of the full range of possible impacts on the natural and socio-economic environment; (2) evaluation and

quantification of these impacts; (3) formulation of measures to avoid, mitigate, and compensate for the anticipated impacts.

Highway systems generate the highest price in human injury and death, as nearly 50 million persons are injured in traffic accidents every year, not including the 1.2 million deaths. Road traffic injury is the single leading cause of unintentional death in the first five decades of human life. Management of safety is a systematic process that strives to reduce the occurrence and severity of traffic accidents. The man/machine interaction with road traffic systems is unstable and poses a challenge to highway safety management. The key for increasing the safety of highway systems is to design, build,  and maintain them to be far more tolerant of the average range of this man/machine interaction with highways. Technological advancements in highway engineering have improved the design, construction, and maintenance methods used over the years. These advancements have allowed for newer highway safety innovations. By ensuring that all situations and opportunities are identified, considered, and implemented as appropriate, they can be evaluated in every phase of highway planning, design, construction, maintenance, and operation to increase the safety of our highway systems.

The most appropriate location, alignment, and shape of a highway are selected during the design stage. Highway design involves the consideration of three major factors (human, vehicular, and roadway) and how these factors interact to provide a safe highway. Human factors include reaction time for braking and steering, visual acuity for traffic signs and signals, and car-following behavior. Vehicle considerations include vehicle size and dynamics that are essential for determining lane width and maximum slopes, and for the selection of design vehicles. Highway engineers design road geometry to ensure stability of vehicles when

 negotiating curves and grades and to provide adequate sight distances for undertaking passing maneuvers along curves on two-lane, two-way roads. In terms of highway engineer design, several aspects are involved.

Geometric Design

Highway and transportation engineers must

meet many safety, service, and performance standards when designing highways for certain site topography. Highway geometric design primarily refers to the visible elements of

the highways. Highway engineers who design the geometry of highways must also consider environmental and social effects of the design on the surrounding infrastructure. There are certain considerations that must be properly addressed in the design process to successfully fit a highway to a site's topography and maintain its safety. Some of these design considerations include: design speed, design traffic volume, number of lanes, Level of Service (LOS), sight distance, cross section lane width and so on.

Materials

The materials used for roadway construction have progressed with time, dating back to the early days of the Roman Empire. Advancements in methods with which these materials are characterized and applied to pavement structural design have accompanied this advancement in materials. There are two major types of pavement surfaces: Portland cement concrete (PCC) and hot-mix asphalt (HMA). Underneath this wearing course are material layers that give structural support for the pavement system. These underlying surfaces may include either the aggregate base and subbase layers, or treated base and subbase layers, and additionally the underlying natural or treated subgrade. These treated layers may be cement-treated, asphalt-treated, or lime-treated for additional support.

Flexible Pavement Design

A flexible, or asphalt, or tarmac pavement typically consists of three or four layers. A flexible pavement's surface layer is constructed of hot-mix asphalt (HMA). With flexible pavement, the highest stress occurs at the surface and the stress decreases as the depth of the pavement increases. Therefore, the highest quality material needs to be used for the surface,

while lower quality materials can be used as the depth of the pavement increases. The term "flexible" is used because of the asphalts ability to bend and deform slightly, then return to its original position as each traffic load is applied and removed. It is possible for these small deformations to become permanent, which can lead to rutting in the wheel path over an extended time.

The service life of a flexible pavement is typically designed in the range of 20 to 30 years. Required thicknesses of each layer of a flexible pavement vary widely depending on the materials used, magnitude, number of repetitions of traffic loads, environmental conditions, and the desired service life of the pavement. Factors such as these are taken into consideration during the design process so that the pavement will last for the designed life without excessive distresses.

Rigid Pavement Design

Rigid pavements are generally used in constructing airports and major highways, such as those in the interstate highway system. In addition, they commonly serve as heavy-duty industrial floor slabs, port and harbor yard pavements, and heavy-vehicle park or terminal pavements. Like flexible pavements, rigid highway pavements are designed as all-weather, long-lasting structures to serve modern day high-speed traffic. Offering high quality riding surfaces for safe vehicular travel, they function as structural layers to distribute vehicular wheel loads in such a manner that the induced stresses transmitted to the subgrade soil are of acceptable magnitudes.

Portland cement concrete (PCC) is the most common material used in the construction of rigid pavement slabs. The reason for its popularity is due to its availability and the economy. Rigid pavements must be designed to endure frequently repeated traffic loadings. The typical designed service life of a rigid pavement is between 30 and 40 years, lasting about twice as long as a flexible pavement.

One major design consideration of rigid pavements is reducing fatigue failure due to the repeated stresses of traffic. Fatigue failure is common among major roads because a typical  highway will experience millions of wheel passes throughout its service life. In addition to design criteria such as traffic loadings, tensile stresses due to thermal energy must also be taken into consideration. As pavement design has progressed, many highway engineers have noted that thermally induced stresses in rigid pavements can be just as intense as those imposed by wheel loadings. Due to the relatively low tensile strength of concrete, thermal stresses are extremely important to the design considerations of rigid pavements.

Flexible Pavement Overlay Design

Over the service life of a flexible pavement, accumulated traffic loads may cause excessive rutting or cracking, inadequate ride quality, or an inadequate skid resistance. These problems can be avoided by adequately maintaining the pavement, but the solution usually has excessive maintenance costs, or the pavement may have an inadequate structural capacity for the projected traffic loads. Throughout a highway's life, its level of serviceability is closely monitored and maintained. One common method used to maintain a highway's level of serviceability is to place an overlay on the pavement's surface. There are three general types of overlay used on flexible pavements: asphalt-concrete overlay, Portland cement concrete overlay, and ultra-thin Portland cement concrete overlay.

Rigid Pavement Overlay Design

Near the end of a rigid pavement's service life, a decision must be made to either fully reconstruct the worn pavement, or construct an overlay layer. Considering an overlay can be constructed on a rigid pavement that has not reached the end of its service life, it is often

more economically attractive to apply overlay layers more frequently. The required overlay thickness for a structurally sound rigid pavement is much smaller than for one that has reached the end of its service life.

Drainage System Design

Designing for proper drainage of highway systems is crucial to their success. Regardless of how well other aspects of a road are designed and constructed, adequate drainage is mandatory

for a road to survive its entire service life. Excess water in the highway structure can inevitably lead to premature failure, even if the failure is not catastrophic. Each highway drainage system is site-specific and can be very complex. Depending on the geography of the region, many methods for proper drainage may not be applicable. The highway engineer must determine which situations a particular design process should be applied, usually a combination of several appropriate methods and materials to direct water away from the structure.

Highway construction is generally preceded by detailed surveys and subgrade preparation. The methods and technology for constructing highways has evolved over time and become increasingly sophisticated. This advancement in technology has raised the level of skill sets required to manage highway construction projects. This skill varies from project to project, depending on factors such as the project's complexity and nature, the contrasts between new construction and reconstruction, and differences between urban region and rural region projects.

There are a number of elements of highway construction which can be broken up into technical and commercial elements of the system. The technical elements include material, material quality, installation techniques and traffic. The latter elements mainly contain five parts, that is, contract understanding, environmental aspects, political aspects, legal aspects and public concerns.

The overall purpose of highway maintenance is to fix defects and preserve the pavement's structure and serviceability. Defects must be defined, understood, and recorded in order to select an appropriate maintenance plan. Defects differ between flexible and rigid pavements. There are four main objectives of highway maintenance: (1) repair of functional pavement defects; (2) extend the functional and structural service life of the pavement; (3) maintain road safety and signage; (4) keep road reserve in acceptable condition. Through routine maintenance practices, highway systems and all of their components can be maintained to their original, as-built condition.

Project management involves the organization and structuring of project activities from inception to completion. Activities could be the construction of infrastructure such as highways and bridges or major and minor maintenance activities related to constructing such infrastructure. The entire project and involved activities must be handled in a professional manner and completed within deadlines and budget. In addition, minimizing social and environmental impacts is essential to successful project management.

Section B Text Exploration

New Words and Expressions

aggregate ['ægrigət; 'ægrigeit]	n.	合计，总计；集合体
	a.	集合的，合计的
alignment [ə'lainmənt]	n.	队列，成直线；校准；结盟；（野外测量、铁路、公路等的）平面图
catastrophic [ˌkætə'strɒfik]	a.	灾难的，毁灭性的，悲惨的
congest [kən'dʒest]	v.	使充血，充塞，充血，拥挤
	n.	曲线，弯曲，曲线图表
drainage ['dreinidʒ]	n.	排水；排水系统；排水面积
intersection [ˌintə'sekʃən]	n.	十字路口；交叉，交集；交叉点

magnitude	['mægnitjuːd]	n.	大小，量级，震级，光度
maneuver	[mə'nuːvə]	n.	机动；演习；策略；调遣
mandatory	['mændətəriː]	a.	强制的，命令的；托管的
		n.	受托者
mitigate	['mitigeit]	v.	使缓和，使减轻，减轻，缓和下来
overlay	[ˌəuvə'lei; 'əuvəlei]	n.	覆盖图，覆盖物
		v.	在表面上铺一薄层，镀
serviceability	[ˌsəːvisə'biliti]	n.	可用性，适用性，使用可靠性，可服务性，可维修性
slab	[slæb]	n.	平板，厚板，厚块，厚片；扁钢坯；混凝土路面
		v.	用石板铺
subgrade	['sʌbgreid]	n.	路基，地基
tarmac	['taːmæk]	n.	柏油碎石路面，铺有柏油碎石的飞机跑道
tensile	['tensail; 'tensəl]	a.	拉力的；可伸长的，可拉长的
topography	[tə'pɔgrəfi]	n.	地势；地形学；地志
vehicular	[viː'hikjulə]	a.	车辆的，用车辆运载的；作为媒介的
vibration	[vai'breiʃən]	n.	振动；犹豫；心灵感应

flexible pavement	柔性路面，沥青路面
geometric design	几何设计，线形设计
hot-mix asphalt (HMA)	热沥青混合物
Portland cement concrete (PCC)	硅酸盐水泥混凝土，普通水泥混凝土
rigid pavement	刚性路面，混凝土路面
service life	使用寿命，使用期限，耐用年限
traffic flow	交通流量
visual acuity	视觉灵敏度
wearing course	磨耗层

Environmental Impact Assessment (EIA)	环境影响评价
Federal-Aid Highway Act	联邦助建高速公路法案
Roman Empire	罗马帝国（指公元前 27 年到公元 476 年的罗马奴隶制国家）
Transport Research Laboratory (TRL)	运输研究实验室

I True and false.

1. The beginning of road construction could be dated to the time of the Romans. (□T □F)
2. The growth of the motor vehicle industry and accompanying economic growth have generated a demand for safer, better performing, more congested highways. (□T □F)
3. The man/machine interaction with road traffic systems is stable and brings no challenge to highway safety management. (□T □F)
4. A tarmac pavement typically consists of three or four layers. (□T □F)
5. There are two general types of overlay used on flexible pavements: asphalt-concrete overlay and Portland cement concrete overlay. (□T □F)

II Choose the best answer according to the text.

1. The first research dedicated to highway engineering was initiated _____.
   A. in the USA
   B. in the United Kingdom
   C. in 1944
   D. in the late 19th to early 20th century

Unit 10　Road Works

2. EIA systematically deals with _____.
   A. identification of the full range of possible impacts on the natural and socio-economic environment
   B. evaluation and quantification of the impacts
   C. formulation of measures to avoid, mitigate, and compensate for the anticipated impacts
   D. all of A, B and C
3. What are two major types of pavement surfaces?
   A. PCC and HMA.
   B. PCC and EIA.
   C. TRL and HMA.
   D. EIA and TRL.
4. The technical elements of highway construction include _____.
   A. human, vehicular and roadway
   B. political, legal and environmental aspects
   C. material, material quality, installation techniques and traffic
   D. roads, bridges and tunnels
5. Which of the following is true about the service life of a pavement?
   A. The service life of a rigid pavement is typically designed in the range of 20 to 30 years.
   B. The typical designed service life of a flexible pavement is between 30 and 40 years.
   C. The service life of all the pavements can be more than 50 years.
   D. The typical designed service life of a rigid pavement can last about twice as long as a flexible pavement.

III　Translation.

1. 设计公路几何形状的公路工程师也必须考虑其设计对周边基础设施的环境影响和社会影响。
   A. Highway engineers who design the geography of highways must also consider environmental and public effects of the design on the surrounding infrastructure.
   B. Highway engineers who design the geometry of highways must also consider environmental and social effects of the design on the surrounding infrastructure.
   C. Highway engineers who design the geography of highways must also consider environmental or social effects of the design on the surrounding infrastructure.
   D. Highway engineers who design the geometry of highways must also consider

either environmental or public effects of the design on the surrounding infrastructure.

2. 发达国家经常面临花高额费用维护老化的交通公路的问题。
   A. Developed countries are constantly faced with high maintenance cost of crumbling transportation railways.
   B. Developing countries are occasionally faced with high maintenance cost of aging transportation highways.
   C. Developed countries are constantly faced with high maintenance cost of aging transportation highways.
   D. Developing countries are occasionally faced with high maintenance cost of aging transportation highways.

3. The entire project and involved activities must be handled in a professional manner and completed within deadlines and budget. In addition, minimizing social and environmental impacts is essential to successful project management.
   A. 整个项目和相关活动必须以专业的方式进行处理，并且不能超出最后期限和预算。此外，使社会影响和环境影响最小化对于成功的项目管理至关重要。
   B. 全部的项目和相关活动必须以专业的方式进行处理，并且不能超出最后期限和预算。此外，使社会影响和环境影响最大化对于成功的项目管理至关重要。
   C. 整个项目和相关活动必须以专业的方式进行处理，并且不能超出最后期限和预算。然而，使社会影响和环境影响最小化对于项目管理的成功至关重要。
   D. 整个项目和其他活动应该以专业的方式进行处理，并且不能超出最后期限和预算。此外，使社交影响和环境作用最大化对于成功的项目管理至关重要。

Section C Supplementary Reading

Road Traffic Safety

Road traffic safety refers to the methods and measures used to prevent road users from being killed or seriously injured. Typical road users include pedestrians, cyclists, motorists, vehicle passengers, and passengers of on-road public transport (mainly buses and trams).

As sustainable solutions for all classes of road have not been identified, particularly low-traffic rural and remote roads, a hierarchy of control should be applied, similar to classifications used to improve occupational safety and health. At the highest level is sustainable prevention of serious injury and death crashes, with sustainable requiring all key result areas to be considered. At the second level is real time risk reduction, which involves providing users at severe risk with a specific warning to enable them to take mitigating action. The third level is about reducing the crash risk which involves applying the road design standards and guidelines, improving driver behavior and enforcement.

Road traffic crashes are one of the world's largest public health and injury prevention problems. The problem is all the more acute because the victims are overwhelmingly healthy before their crashes. According to the World Health Organization (WHO), more than 1 million people are killed on the world's roads each year. Report published by the WHO in 2004 estimated that some 1.2 million people were killed and 50 million injured in traffic collisions on the roads around the world each year and was the leading cause of death among children 10-19 years of age. The report also noted that the problem was most severe in developing countries and that simple prevention measures could halve the number of deaths.

The standard measures used in assessing road safety interventions are fatalities and killed or seriously injured (KSI) rates, usually per billion (10^9) passenger kilometers. Countries caught in the old road safety paradigm, replace KSI rates with crash rates—for example, crashes per million vehicle miles.

Vehicle speed within the human tolerances for avoiding serious injury and death is a key goal of modern road design because impact speed affects the severity of injury to both occupants and pedestrians. For occupants, Joksch (1993) found the probability of death for drivers in multi-vehicle accidents increased as the fourth power of impact speed. Injuries are caused by sudden, severe acceleration (or deceleration); this is difficult to measure. However, crash reconstruction techniques can estimate vehicle speeds before a crash. Therefore, the change in speed is used as a surrogate for acceleration. This enabled the Swedish Road Administration to identify the KSI risk curves using actual crash reconstruction data which led to the human tolerances for serious injury and

death referenced above.

On neighborhood roads where many vulnerable road users, such as pedestrians and bicyclists can be found, traffic calming can be a tool for road safety. Shared space schemes, which rely on human instincts and interactions, such as eye contact, for their effectiveness, and are characterized by the removal of traditional traffic signals and signs, and even by the removal of the distinction between carriageway (roadway) and footway (sidewalk), are also becoming increasingly popular. Both approaches can be shown to be effective.

Modern safety barriers are designed to absorb impact energy and minimize the risk to the occupants of cars and bystanders. For example, most side rails are now anchored to the ground, so that they cannot skewer a passenger compartment. Most light poles are designed to break at the base rather than violently stop a car that hits them. Some road fixtures such as signs and fire hydrants are designed to collapse on impact. Highway authorities have removed trees in the vicinity of roads; while the idea of "dangerous trees" has attracted a certain amount of skepticism, unforgiving objects such as trees can cause severe damage and injury to errant road users. Safety barriers can provide some combination of physical protection and visual protection depending on their environment. Physical protection is important for protecting sensitive building and pedestrian areas. Visual protection is necessary to alert drivers to changes in road patterns.

Most roads are cambered (crowned), that is, made so that they have rounded surfaces, to reduce standing water and ice, primarily to prevent frost damage but also increasing traction in poor weather. Some sections of road are now surfaced with porous bitumen to enhance drainage; this is particularly done on bends. These are just a few elements of highway engineering. As well as that, there are often grooves cut into the surface of cement highways to channel water away, and rumble strips at the edges of highways to rouse inattentive drivers with the loud noise they make  when driven over. In some cases, there are raised markers between lanes to reinforce the lane boundaries; these are often reflective. In pedestrian areas, speed bumps are often placed to slow cars, preventing them from going too fast near pedestrians.

Poor road surfaces can lead to safety problems. If too much asphalt or bituminous binder is used in asphalt concrete, the binder can "bleed or flush" to the surface, leaving a very smooth surface that provides little traction when wet. Certain kinds of stone aggregate become very smooth or polished under the constant wearing action of vehicle tyres, again leading to poor wet-weather traction. Either  of these problems can increase wet-weather crashes by increasing braking distances or contributing to loss of control. If the pavement is insufficiently sloped or poorly drained, standing water on the surface can also lead to wet-weather crashes due to hydroplaning.

Lane markers in some countries and states are marked with cat's eyes, Botts' dots or reflective raised pavement markers that do not fade like paint. Botts' dots are not used where it is icy in the winter, because frost and snowplows can break the glue that holds them to the road, although they can be embedded in short, shallow trenches carved in the roadway, as is done in the mountainous regions of California.

Road hazards and intersections in some areas are now usually marked several times, roughly five, twenty, and sixty seconds in advance so that drivers are less likely to attempt violent manoeuvres. Most road signs and pavement marking materials are retro-reflective, incorporating small glass spheres or prisms to more efficiently reflect light from vehicle headlights back to the driver's eyes.

Major highways including motorways, freeways, autobahnen and interstates are designed for safer high-speed operation and generally have lower levels of injury per vehicle km than other roads; for example, in 2013, the German autobahn fatality rate of 1.9 deaths per billion-travel-kilometers compared favorably with the 4.7 rate on urban streets and 6.6 rate on rural roads. Safety features include: limited access from the properties and local roads; grade separated junctions; median dividers between opposite-direction traffic to reduce likelihood of head-on collisions; removing roadside obstacles; prohibition of more vulnerable road users; placements of energy attenuation devices (e.g. guard rails, wide grassy areas, sand barrels); and eliminating road toll booths.

Safety can be improved in various ways depending on the transport taken.

Buses and Coaches

Safety can be improved in various simple ways to reduce the chance of an accident occurring. Avoiding rushing or standing in unsafe places on the bus or coach and following the rules on the bus or coach itself will greatly increase the safety of a person travelling by bus or coach. Various safety features can also be implemented into buses and coaches to improve safety including safety bars for people to hold onto. The main ways to stay safe when travelling by bus or coach are as follows:

(1) Leave your location early so that you do not have to run to catch the bus or coach.

(2) At the bus stop, always follow the queue.

(3) Do not board or alight at a bus stop other than an official one.

(4) Never board or alight at a red light crossing or unauthorized bus stop.

(5) Board the bus only after it has come to a halt without rushing in or pushing others.

(6) Do not sit, stand or travel on the footboard of the bus.

(7) Do not put any part of your body outside a moving or a stationary bus.

(8) While in the bus, refrain from shouting or making noise as it can distract the driver.

(9) Always hold onto the handrail if standing in a moving bus, especially on sharp turns.

(10) Always adhere to the bus safety rules.

Cars

Safety can be improved by reducing the chances of a driver making an error, or by designing vehicles to reduce the severity of crashes that do occur. Most industrialized countries have comprehensive requirements and specifications for safety-related vehicle devices, systems, design, and construction. These may include:

(1) Passenger restraints such as seat belts—often in conjunction with laws requiring their use—and airbags.

(2) Crash avoidance equipment such as lights and reflectors.

(3) Driver assistance systems such as Electronic Stability Control.

(4) Crash survivability design including fire-retardant interior materials, standards for fuel system integrity, and the use of safety glass.

(5) Sobriety detectors. These interlocks prevent the ignition key from working if the driver breathes into one and it detects significant quantities of alcohol. They have been used by some commercial transport companies, or suggested for use with persistent drunk-driving offenders on a voluntary basis.

Motorbikes

UK road casualty statistics show that motorcycle riders are nine times more likely to crash, and 17 times more likely to die in a crash, than car drivers. The higher fatality risk is due in part to the lack of crash protection (unlike in enclosed vehicles such as cars), combined with the high speeds motorcycles typically travel at. According to the US statistics, the percentage of intoxicated motorcyclists in fatal crashes is higher than other riders on roads. Helmets also play a major role in the safety of motorcyclists. In 2008, the National Highway Traffic Safety Administration (NHTSA) estimated the helmets are 37 percent effective in saving lives of motorcyclists involved in crashes.

Trucks

According to the European Commission Transportation Department, "It has been estimated that up to 25% of accidents involving trucks can be attributable to inadequate cargo securing". Improperly-secured cargo can cause severe accidents and lead to loss of cargo, loss of lives, loss of vehicles, and can be a hazard for the environment. One way to stabilize, secure, and protect cargo during transportation on the road is by using dunnage bags, which are placed in the voids among the cargo and are designed to prevent the load from moving during transport.

Together for Safer Roads (TSR) has developed best practices for implementing corporate road safety programs that includes data management and analysis, route mapping, investment and upkeep of fleets, safety policies and training for employees, and first-aid/safety training in case collisions do occur.

Police

Hundreds of people are killed each year due to high-speed chases of fleeing suspects by police. Different jurisdictions allow such pursuits in different circumstances; fewer injuries might occur if these are restricted to violent felonies.

Unit 11
Concrete Works

Section A Text

Reinforced Concrete

Reinforced concrete (RC) is a composite material in which concrete's relatively low tensile strength and ductility are counteracted by the inclusion of reinforcement having higher tensile strength or ductility. The reinforcement is usually, though not necessarily, a steel reinforcing bar (rebar) and is usually embedded passively in the concrete before the concrete sets. Reinforcing schemes are generally designed to resist tensile stresses in particular regions of the concrete that might cause unacceptable cracking and/or structural failure. Modern reinforced concrete can contain  varied reinforcing materials made of steel, polymers or alternate composite material in conjunction with rebar or not. Reinforced concrete may also be permanently stressed (in tension), so as to improve the behavior of the final structure under working loads. In the United States, the most common methods of doing this are known as pre-tensioning and post-tensioning.

For a strong, ductile and durable construction the reinforcement needs to have the following properties at least:

(1) High relative strength.

(2) High toleration of tensile strain.

(3) Good bond to the concrete, irrespective of pH, moisture, and similar factors.

(4) Thermal compatibility, not causing unacceptable stresses in response to changing temperatures.

(5) Durability in the concrete environment, irrespective of corrosion or sustained stress for example.

Use in Construction

Many different types of structures and components of structures can be built using reinforced concrete including slabs, walls, beams, columns, foundations, frames and more.

Reinforced concrete can be classified as precast or cast-in-place concrete.

Designing and implementing the most efficient floor system is key to creating optimal building structures. Small changes in the design of a floor system can have significant impact on material costs, construction schedule, ultimate strength, operating costs, occupancy levels and end use of a building.

Without reinforcement, constructing modern structures with concrete material would not be possible.

Behavior of Reinforced Concrete

1. Materials

Concrete is a mixture of coarse (stone or brick chips) and fine (generally sand or crushed stone) aggregates with a paste of binder material (usually Portland cement) and water. When cement is mixed with a small amount of water, it hydrates to form microscopic opaque crystal lattices encapsulating and locking the aggregate into a rigid structure. The aggregates used for making concrete should be free from harmful substances like organic impurities, silt, clay, lignite etc. Typical concrete mixes have high

resistance to compressive stresses; however, any appreciable tension (e. g., due to bending) will break the microscopic rigid lattice, resulting in cracking and separation of the concrete. For this reason, typical non-reinforced concrete must be well supported to prevent the development of tension.

If a material with high strength in tension, such as steel, is placed in concrete, then the composite material, reinforced concrete, resists not only compression but also bending and other direct tensile actions. A reinforced concrete section where the concrete resists the compression and steel resists the tension can be made into almost any shape and size for the construction industry.

2. Key characteristics

Three physical characteristics give reinforced concrete its special properties.

(1) The coefficient of thermal expansion of concrete is similar to that of steel, eliminating large internal stresses due to differences in thermal expansion or contraction.

(2) When the cement paste within the concrete hardens, this conforms to the surface details of the steel, permitting any stress to be transmitted efficiently between the different materials. Usually steel bars are roughened or corrugated to further improve the bond or cohesion between the concrete and steel.

(3) The alkaline chemical environment provided by the alkali reserve (KOH, NaOH) and the portlandite (calcium hydroxide) contained in the hardened cement paste causes a passivating film to form on the surface of the steel, making it much more resistant to corrosion than it would be in neutral or acidic conditions. When the cement paste is exposed to the air and meteoric water reacts with the atmospheric CO_2, portlandite and the calcium silicate hydrate (CSH) of the hardened cement paste become progressively carbonated and the high pH gradually decreases from 13.5-12.5 to 8.5, the pH of water in equilibrium with calcite (calcium carbonate) and the steel is no longer passivated.

3. Anchorage (bond) in concrete: codes of specifications

Because the actual bond stress varies along the length of a bar anchored in a zone of tension, current international codes of specifications use the concept of development length rather than bond stress. The main requirement for safety against bond failure is to provide a sufficient extension of the length of the bar beyond the point where the steel is required to develop its yield stress and this length must be at least equal to its development length. However, if the actual available length is inadequate for full development, special anchorages must be provided, such as cogs or hooks.

Reinforcement and Terminology of Beams

A beam bends under bending moment, resulting in a small curvature. At the outer face (tensile face) of the curvature the concrete experiences tensile stress, while at the inner face (compressive face) it experiences compressive stress.

A singly reinforced beam is one in which the concrete element is only reinforced near the tensile face and the reinforcement, called tension steel, is designed to resist the tension.

A doubly reinforced beam is one in which besides the tensile reinforcement the concrete element is also reinforced near the compressive face to help the concrete resist compression. The latter reinforcement is called compression steel. When the compression zone of a concrete is inadequate to resist the compressive moment (positive moment), extra reinforcement has to be provided if the architect limits the dimensions of the section.

An under-reinforced beam is one in which the tension capacity of the tensile reinforcement is smaller than the combined compression capacity of the concrete and the compression steel (under-reinforced at tensile face). When the reinforced concrete element is subject to increasing bending moment, the tension steel yields while the concrete does not reach its

ultimate failure condition. As the tension steel yields and stretches, an "under-reinforced" concrete also yields in a ductile manner, exhibiting a large deformation and warning before its ultimate failure. In this case the yield stress of the steel governs the design.

An over-reinforced beam is one in which the tension capacity of the tension steel is greater than the combined compression capacity of the concrete and the compression steel (over-reinforced at tensile face). So the "over-reinforced concrete" beam fails by crushing of the compressive-zone concrete and before the tension zone steel yields, which does not provide any warning before failure as the failure is instantaneous.

A balanced-reinforced beam is one in which both the compressive and tensile zones reach yielding at the same imposed load on the beam, and the concrete will crush and the tensile steel will yield at the same time. This design criterion is however as risky as over-reinforced concrete, because failure is sudden as the concrete crushes at the same time of the tensile steel yields, which gives a very little warning of distress in tension failure.

Steel-reinforced concrete moment-carrying elements should normally be designed to be under-reinforced so that users of the structure will receive warning of impending collapse.

The characteristic strength is the strength of a material where less than 5% of the specimen shows lower strength.

The design strength or nominal strength is the strength of a material, including a material-safety factor. The value of the safety factor generally ranges from 0.75 to 0.85 in permissible stress design.

The ultimate limit state is the theoretical failure point with a certain probability. It is stated under factored loads and factored resistances.

Prestressed Concrete

Prestressing concrete is a technique that greatly increases the load-bearing strength of concrete beams. The reinforcing steel in the

bottom part of the beam, which will be subjected to tensile forces when in service, is placed in tension before the concrete is poured around it. Once the concrete has hardened, the tension on the reinforcing steel is released, placing a built-in compressive force on the concrete. When loads are applied, the reinforcing steel takes on more stress and the compressive force in the concrete is reduced, but does not become a tensile force. Since the concrete is always under compression, it is less subject to cracking and failure.

Common Failure Modes of Steel Reinforced Concrete

Reinforced concrete can fail due to inadequate strength, leading to mechanical failure, or due to a reduction in its durability. Corrosion and freeze/thaw cycles may damage poorly designed or constructed reinforced concrete. When rebar corrodes, the oxidation products (rust) expand and tends to flake, cracking the concrete and unbonding the rebar from the concrete. Typical mechanisms leading to durability problems are discussed below.

1. Mechanical failure

Cracking of the concrete section is nearly impossible to prevent; however, the size and location of cracks can be limited and controlled by appropriate reinforcement, control joints, curing methodology and concrete mix design. Cracking can allow moisture to penetrate and corrode the reinforcement. This is a serviceability failure in limit state design. Cracking is normally the result of an inadequate quantity of rebar, or rebar spaced at too great a distance. The concrete then cracks either under excess loading, or due to internal effects such as early thermal shrinkage while it cures.

Ultimate failure leading to collapse can be caused by crushing the concrete, which occurs when compressive stresses exceed its strength, by yielding or failure of the rebar when bending or shear stresses exceed the strength of the reinforcement, or by bond failure between the concrete and the rebar.

2. Carbonation

Carbonation, or neutralisation, is a chemical reaction between carbon dioxide in the air and calcium hydroxide and hydrated calcium silicate in the concrete.

When a concrete structure is designed, it is usual to specify the concrete cover for the rebar (the depth of the rebar within the object). The minimum concrete cover is normally regulated by design or building codes. If the reinforcement is too close to the surface, early failure due to corrosion may occur. The concrete cover depth can be measured with a cover meter. However, carbonated concrete incurs a durability problem only when there is also sufficient moisture and oxygen to cause electropotential corrosion of the reinforcing steel.

One method of testing a structure for carbonatation is to drill a fresh hole in the surface and then treat the cut surface with phenolphthalein indicator solution. This solution turns pink when in contact with alkaline concrete, making it possible to see the depth of carbonation. Using an existing hole does not suffice because the exposed surface will already be carbonated.

3. Conversion of high alumina cement

Resistant to weak acids and especially sulfates, this cement cures quickly and has very high durability and strength. It was frequently used after World War II to make precast concrete objects. However, it can lose strength with heat or time (conversion), especially when not properly cured. After the collapse of three roofs made of prestressed concrete beams using high alumina cement, this cement was banned in the UK in 1976. Subsequent inquiries into the matter showed that the beams were improperly manufactured, but the ban remained.

Section B Text Exploration

New Words and Expressions

aggregate ['ægrigət]	n.	（混凝土或灰泥的）骨料，集料，粒料
alkaline ['ælkəlain]	a.	碱的，似碱的
anchorage ['æŋkəridʒ]	n.	加固物，锚具
carbonation [ˌkɑːbə'neiʃən]	n.	碳酸饱和，碳化作用
collapse [kə'læps]	n.	倒塌
contraction [kən'trækʃən]	n.	收缩性；收缩，缩小
corrode [kə'rəud]	v.	腐蚀，侵蚀；损伤；使恶化
curvature ['kəːvətʃə]	n.	弯曲；弯曲物
ductility [dʌk'tiləti]	n.	延（展）性，可延展性，可锻性，可塑性，塑性，韧性
encapsulate [in'kæpsəleit]	v.	包于囊中，用胶囊（或囊状物）包（或装）；压缩，缩短
exceed [ik'siːd]	v.	越出，超越
flake [fleik]	v.	使成薄片，把……切成薄片；使成片剥落；剥，削，凿；像雪花般覆盖；雪片似的降落
hydrate ['haidreit]	v.	与水化合
lattice ['lætis]	n.	格子，方格，格栅，格架，支承桁架
lignite ['lignait]	n.	褐煤
linear ['liniə]	a.	线的，直线的；线构的
microscopic [ˌmaikrə'skɔpik]	a.	用显微镜可见的，微小的，细小的，微观的，极小的
moisture ['mɔistʃə]	n.	潮湿，湿气，潮气；湿度；水气，水分

neutralisation	[ˌnjuːtrəlaiˈzeiʃən]	n.	中和，中立状态
opaque	[əuˈpeik]	a.	不透明的，不透光的，（对声音、热、辐射等）不传导的
optimal	[ˈɔptiməl]	a.	最适宜的，最佳的，最优秀的
oxidation	[ˌɔksiˈdeiʃən]	n.	氧化，氧化作用
passivate	[ˈpæsiveit]	v.	使钝化
penetrate	[ˈpenitreit]	v.	渗透，浸入
phenolphthalein	[ˌfiːnɔlˈfθæliin]	n.	（苯）酚酞（用作酸碱滴定指示剂及轻泻剂）
polymer	[ˈpɔlimə]	n.	聚合物，聚合体
portlandite	[ˈpɔːtləndait]	n.	氢氧钙石，羟钙石
release	[riˈliːs]	v.	释放，放出，排放
silt	[silt]	n.	泥沙，淤泥
sulfate	[ˈsʌlfeit]	n.	硫酸盐
terminology	[ˌtəːmiˈnɔlədʒi]	n.	术语，专有名词，术语学

be embedded in	嵌入……
be mixed with	与……混合在一起
be subject to	可能受……影响的，易遭受……的
binder material	黏合剂，黏合料
bond stress	（混凝土与钢筋的）握裹力，黏结力
calcium carbonate	碳酸钙
calcium hydroxide	氢氧化钙，氢化钙
calcium silicate hydrate (CSH)	水化硅酸钙
cast-in-place concrete	现浇混凝土
conform to	与……相符合，与……相一致；遵守……，服从……
freeze thaw cycles	冰融循环
high alumina cement	高铝水泥
KOH	氢氧化钾
NaOH	氢氧化钠
thermal compatibility	热兼容

I True and false.

1. Reinforced concrete may also be permanently stressed (in tension), so as to improve the behavior of the final structure under working loads. (□T □F)
2. Without reinforcement, constructing modern structures with concrete material would not be possible. (□T □F)
3. The aggregates used for making concrete should contain organic impurities, silt, clay, lignite, etc. (□T □F)
4. The nominal strength is the strength of a material where less than 5% of the specimen shows lower strength. (□T □F)
5. Reinforced concrete can fail due to inadequate strength, leading to mechanical failure, or due to a reduction in its durability. (□T □F)

II Choose the best answer according to the text.

1. Small changes in the design of a floor system can have significant impact on the following aspects except _____.
   A. material costs
   B. construction schedule
   C. operating costs
   D. fire resistance of the building structure
2. Which one is NOT true about the physical characteristics of reinforced concrete?
   A. The coefficient of thermal expansion of concrete is similar to that of steel, eliminating large internal stresses due to differences in thermal expansion or contraction.
   B. When the cement paste within the concrete hardens, this conforms to the surface details of the steel, permitting any stress to be transmitted efficiently between the different materials.
   C. The alkaline chemical environment provided by the alkali reserve (KOH, NaOH) and the portlandite (calcium hydroxide) contained in the hardened

cement paste causes a passivating film to form on the surface of the steel, making it less resistant to corrosion than it would be in neutral or acidic conditions.

D. The alkaline chemical environment provided by the alkali reserve (KOH, NaOH) and the portlandite (calcium hydroxide) contained in the hardened cement paste causes a passivating film to form on the surface of the steel, making it much more resistant to corrosion than it would be in neutral or acidic conditions.

3. Because the actual bond stress varies along the length of a bar anchored in a zone of tension, current international codes of specifications use the concept of _____.
   A. development length
   B. bond stress
   C. tensile stress
   D. compressive stress

4. Which one is true about the balanced-reinforced beam according to the text?
   A. It is one in which besides the tensile reinforcement the concrete element is also reinforced near the compressive face to help the concrete resist compression.
   B. It is one in which the tension capacity of the tensile reinforcement is smaller than the combined compression capacity of the concrete and the compression steel (under-reinforced at tensile face).
   C. It is one in which the tension capacity of the tension steel is greater than the combined compression capacity of the concrete and the compression steel (over-reinforced at tensile face).
   D. It is one in which both the compressive and tensile zones reach yielding at the same imposed load on the beam, and the concrete will crush and the tensile steel will yield at the same time.

5. _____ is a technique that greatly increases the load-bearing strength of concrete beams.
   A. Reinforced concrete
   B. Prestressing concrete
   C. Concrete
   D. Plain concrete

III Translation.

1. 钢筋混凝土截面结合了混凝土抗压和钢筋抗拉的特性，在建筑行业中几乎可以被制成任何的形状和尺寸。
   A. A reinforced concrete section where the concrete resists the compression and steel resists the tension can be made into almost any shape and size for the construction industry.
   B. A reinforced concrete section where the concrete resists the tension and steel resists the compression can be made into any shape and size for the construction industry.
   C. A reinforced concrete section where the steel resists the bending and concrete resists the tension can be made into different shapes and sizes for the construction industry.
   D. A reinforced concrete section where the steel resists the compression and concrete resists the corrosion can be made into different shapes and sizes for the construction industry.

2. 混凝土截面的开裂几乎是不可避免的，但是可以通过适当的加固、接缝控制、养护方法及混凝土配合比设计来限定和控制裂缝的大小和位置。
   A. Cracking of high alumina cement is nearly impossible to prevent; however, the size and location of cracks can be limited and controlled by appropriate reinforcement, control joints, curing methodology and carbonation.
   B. Cracking of high alumina cement is impossible to prevent; however, the size and depth of cracks can be limited and controlled by appropriate reinforcement, control joints, curing methodology and concrete mix design.
   C. Cracking of the concrete section is impossible to prevent; however, the size and depth of cracks can be limited and controlled by appropriate reinforcement, control joints, curing methodology and carbonation.
   D. Cracking of the concrete section is nearly impossible to prevent; however, the size and location of cracks can be limited and controlled by appropriate reinforcement, control joints, curing methodology and concrete mix design.

Section C Supplementary Reading

Environmental Impact of Concrete

The environmental impact of concrete, its manufacture and applications, is complex. Some effects are harmful; others welcome. Many depend on circumstances. A major component of concrete is cement, which has its own environmental and social impacts and contributes largely to those of concrete.

The cement industry is one of the primary producers of carbon dioxide, a major greenhouse gas. Concrete causes damage to the most fertile layer of the earth, the topsoil. Concrete is used to create hard surfaces which contribute to surface runoff that may cause soil erosion, water pollution and flooding. Conversely, concrete is one of the most powerful tools for proper flood control, by means of damming, diversion, and deflection of flood waters, mud flows, and the like. Light-colored concrete can reduce the urban heat island effect, due to its higher albedo. Concrete dust  released by building demolition and natural disasters can be a major source of dangerous air pollution. The presence of some substances in concrete, including useful and unwanted additives, can cause health concerns due to toxicity and radioactivity. Wet concrete is highly alkaline and should always be handled with proper protective equipment. Concrete recycling is increasing in response to improved environmental awareness, legislation, and economic considerations.

Design Improvements

The concrete industry is one of two largest producers of carbon dioxide (CO_2), creating up to 5% of worldwide man-made emissions of this gas, of which 50% is from the chemical process and 40% from burning fuel. There is a growing interest in reducing carbon emissions related to concrete from both the academic and industrial sectors, especially with the possibility of future carbon tax implementation. Several approaches to reducing emissions have been suggested.

1. Cement production and use

One reason why the carbon emissions are so high is because cement has to be heated to very high temperatures in order for clinker to form. A major culprit of this is alite (Ca_3SiO_5), a mineral in concrete that cures within hours of pouring and is therefore responsible for much of its initial strength. However, alite also has to be heated to 1,500 °C in the clinker-forming process. Some research suggests that alite can be replaced by a different mineral, such as belite (Ca_2SiO_4). Belite is also a mineral already used in concrete. It has a roasting temperature of 1,200 °C, which is significantly lower than that of alite. Furthermore, belite is actually stronger once concrete cures. However, belite takes on the order of days or months to set completely, which leaves concrete weak for an unacceptably long time. Current research is focusing on finding possible impurity additives, like magnesium, that might speed up the curing process. It is also worthwhile to consider that belite takes more energy to grind, which may make its full life impact similar to or even higher than alite.

Another approach has been the partial replacement of conventional clinker with such alternatives as fly ash, bottom ash, and slag, all of which are by-products of other industries that would otherwise end up in landfills. Fly ash and bottom ash come from thermoelectric power plants, while slag is a waste from blast furnaces in the ironworks industry. These materials are slowly gaining popularity as additives, especially since they can potentially increase strength, decrease density, and prolong durability of concrete.

The main obstacle to wider implementation of fly ash and slag may be largely due to the risk of construction with new technology that has not been exposed to long field testing. Until a carbon tax is implemented, companies are unwilling to take the chance with new concrete mix recipes even if this reduces carbon emissions. However, there are some examples of "green" concrete and its implementation.

One instance is a concrete company called Ceratech that has started manufacturing concrete with 95% fly ash and 5% liquid additives. Another is the I-35W Saint Anthony Falls Bridge, which was constructed with a novel mixture of concrete that included different compositions of Portland cement, fly ash, and slag depending on the portion of the bridge and its material properties requirements.

2. Emission absorbing concrete

Italian company Italcementi designed a kind of cement that is supposed to fight air pollution. It should break down pollutants that come in contact with the concrete, thanks to the use of titanium dioxide absorbing ultraviolet light. Some environmental experts nevertheless remain skeptical and wonder if the special material can "eat" enough pollutants to make it financially viable. Jubilee Church in Rome is built from this kind of concrete.

Another proposed method of absorbing emissions is to absorb CO_2 in the curing process. Recent research has proposed the use of an admixture (a dicalcium silicate γ phase) that absorbs CO_2 as the concrete cures. With the use of coal ash or another suitable substitute, this concrete could theoretically have a CO_2 emissions below 0 kg/m^3, compared to normal concrete at 400 kg/m^3. The most effective method of production of this concrete would be using the exhaust gas of a power plant, where an isolated chamber could control temperature and humidity. Even besides the use of advanced additives, carbonation naturally occurs within concrete, thus causing it to absorb CO_2 in a process that is effectively the reverse of cement production. While concerns about corrosion of reinforcement and alkalinity loss remain, this process cannot be discounted.

3. Other improvements

There are many other improvements to concrete that do not deal directly with emissions. Recently, much research has gone into "smart" concretes: concretes that use electrical and mechanical signals to respond to changes in loading conditions. One variety uses carbon fiber reinforcement which provides an electrical response that can be used to measure strain. This allows for monitoring the structural integrity of the concrete without installing sensors.

The road construction and maintenance industry consumes tonnes of carbon intensive concrete every day to secure road-side and urban infrastructure. As populations grow this infrastructure is becoming increasingly vulnerable to impact from vehicles, creating an ever increasing cycle of damage and waste and ever increasing consumption of concrete for repairs (roadworks are now seen around our cities on almost a daily basis). A major development in the infrastructure industry involves the use of recycled petroleum waste to protect the concrete from damage and enable infrastructure to become dynamic, able to be easily maintained and updated without disturbance to the existing foundations. This simple innovation preserves the foundations for the entire lifespan of a development.

Another area of concrete research involves the creation of certain "waterless" concretes for use in extraplanetary colonization. Most commonly, these concretes use sulfur to act as a non-reactive binder, allowing for construction of concrete structures in environments with no or very little water. These concretes are in many ways indistinguishable from normal hydraulic concrete: they have similar densities, can be used with currently existing metal reinforcement, and they actually gain strength faster than normal concrete. This application has yet to be explored on Earth, but with concrete production representing as much as two-thirds of the total energy usage of some developing countries, any improvement is worth considering.

Carbon Concrete

Another approach is to pump liquid carbon dioxide into the concrete before mixing. This can reduce the carbon emissions from concrete production when combined with a power plant or other industry that produces CO_2.

Surface Runoff

Surface runoff, when water runs off impervious

surfaces, such as non-porous concrete, can cause severe soil erosion and flooding. Urban runoff tends to pick up gasoline, motor oil, heavy metals, trash and other pollutants from sidewalks, roadways and parking lots. Without attenuation, the impervious cover in a typical urban area limits groundwater percolation and causes five times the amount of runoff generated by a typical woodland of the same size.

In an attempt to counteract the negative effects of impervious concrete, many new paving projects have begun to use pervious concrete, which provides a level of automatic stormwater management. Pervious concrete is created by careful laying of concrete with specifically designed aggregate proportions, which allows for surface runoff to seep through and return to the groundwater. This both prevents flooding and contributes to groundwater replenishment. If designed and layered properly, pervious concrete and other discreetly paved areas can also function as an automatic water filter by preventing certain harmful substances like oils and other chemicals from passing through. Unfortunately there are still downsides to large scale applications of pervious concrete: its reduced strength relative to conventional concrete limits use to low-load areas, and it must be laid properly to reduce susceptibility to freeze-thaw damage and sediment buildup.

Urban Heat

Both concrete and asphalt are the primary contributors to what is known as the urban heat island effect.

Using light-colored concrete has proven effective in reflecting up to 50% more light than asphalt and reducing ambient temperature. A low albedo value, characteristic of black asphalt, absorbs a large percentage of solar heat and contributes to the warming of cities. By paving with light colored concrete, in addition to replacing asphalt with light-colored concrete, communities can lower their average temperature.

In many US cities, pavement covers about 30%-40% of the surface area. This directly affects the temperature of the city and contributes to the urban heat island effect. Paving with

light-colored concrete would lower temperatures of paved areas and improve night-time visibility. The potential of energy saving within an area is also high. With lower temperatures, the demand for air conditioning theoretically decreases, saving energy. However, research into the interaction between reflective pavements and buildings has found that, unless the nearby buildings are fitted with reflective glass, solar radiation reflected off pavements can increase building temperatures, increasing air conditioning demands.

Concrete Dust

Building demolition and natural disasters such as earthquakes often release a large amount of concrete dust into the local atmosphere. Concrete dust was concluded to be the major source of dangerous air pollution following the Great Hanshin earthquake.

Toxic and Radioactive Contamination

The presence of some substances in concrete, including useful and unwanted additives, can cause health concerns. Natural radioactive elements (K, U, Th, and Rn) can be present in various concentration in concrete dwellings, depending on the source of the raw materials used. For example, some stones naturally emit Radon, and Uranium was once common in mine refuse. Toxic substances may also be unintentionally used as the result of contamination from a nuclear accident. Dust from rubble or broken concrete upon demolition or crumbling may cause serious health concerns depending also on what had been incorporated in the concrete. However, embedding harmful materials in concrete is not always dangerous and may in fact be beneficial. In some cases, incorporating certain compounds such as metals in the hydration process of cement immobilizes them in a harmless state and prevents them from being released freely elsewhere.

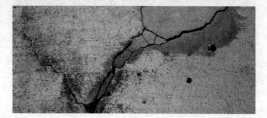

Handling Precautions

Handling of wet concrete must always be

done with proper protective equipment. Contact with wet concrete can cause skin chemical burns due to the caustic nature of the mixture of cement and water. Indeed, the pH of fresh cement water is highly alkaline due to the presence of free potassium and sodium hydroxides in solution. Eyes, hands and feet must be correctly protected to avoid any direct contact with wet concrete and washed without delay if necessary.

Concrete Recycling

Concrete recycling is an increasingly common method of disposing of concrete structures. Concrete debris was once routinely shipped to landfills for disposal, but recycling is increasing due to improved environmental awareness, governmental laws and economic benefits.

Concrete, which must be free of trash, wood, paper and other such materials, is collected from demolition sites and put through a crushing machine, often along with asphalt, bricks and rocks.

Reinforced concrete contains rebar and other metallic reinforcements, which are removed with magnets and recycled elsewhere. The remaining aggregate chunks are sorted by size. Larger chunks may go through the crusher again. Smaller pieces of concrete are used as gravel for new construction projects. Aggregate base gravel is laid down as the lowest layer in a road, with fresh concrete or asphalt placed over it. Crushed recycled concrete can sometimes be used as the dry aggregate for brand new concrete if it is free of contaminants, though the use of recycled concrete limits strength and is not allowed in many jurisdictions.

Unit 12
Steel Works

Section A Text

Structural Steel

Structural steel is a category of steel used as a construction material for making structural steel shapes. A structural steel shape is a profile, formed with a specific cross section and following certain standards for chemical composition and mechanical properties. Structural steel shapes, sizes, composition, strengths, storage practices, etc., are regulated by standards in most industrialized countries.

Structural steel members, such as I-beams, have high second moments of area, which allow them to be very stiff in respect to their cross-sectional area.

Common Structural Shapes

The shapes available are described in many published standards worldwide, and a number of specialist and proprietary cross sections are also available.

A steel I-beam, in this case used to support timber joists in a house. I-beam [I-shaped cross section—in Britain these include Universal Beams (UB) and Universal Columns

(UC); in Europe it includes the IPE, HE, HL, HD and other sections; in the US it includes Wide Flange (WF or W-Shape) and H sections].

Z-Shape (half a flange in opposite directions).

HSS-shape [Hollow structural section also known as SHS (structural hollow section) and including square, rectangular, circular (pipe) and elliptical cross sections].

Angle (L-shaped cross section).

Structural channel, or C-beam, or C cross section.

Tee (T-shaped cross section).

Rail profile (asymmetrical I-beam).

Bar, a piece of metal, rectangular cross-sectioned (flat) and long, but not so wide so as to be called a sheet.

Rod, a round or square and long piece of metal.

Plate, metal sheets thicker than 6 mm or 1/4 in.

Open web steel joist.

Standards

1. Standard structural steels (Europe)

Most steels used throughout Europe are specified to comply with the European standard EN 10025. However, many national standards also remain in force.

Typical grades are described as "S275J2" or "S355-K2W". In these examples, "S" denotes structural rather than engineering steel; 275 or 355 denotes the yield strength in newtons per square millimetre or the equivalent megapascals; J2 or K2 denotes the materials toughness by reference to Charpy impact test values; and the "W" denotes weathering steel. Further letters can be used to designate fine grain steel ("N" or "NL"); quenched and tempered steel ("Q" or "QL"); and thermomechanically rolled steel ("M" or "ML").

2. Standard structural steels (USA)

Steels used for building construction in the US use standard alloys identified and specified by ASTM International. These steels have an alloy identification beginning with A and then two, three, or four numbers. The four-number AISI steel grades commonly used for mechanical engineering, machines, and vehicles are a completely different specification series.

3. CE marking

The concept of CE marking for all construction products and steel products is introduced by the Construction Products Directive (CPD). The CPD is a European Directive that ensures the free movement of all construction products within the European Union.

Because steel components are "safety critical", CE Marking is not allowed unless the Factory Production Control (FPC) system under which they are produced has been assessed by a suitable certification body that has been approved to the European Commission.

In the case of steel products such as sections, bolts and fabricated steelwork the CE Marking demonstrates that the product complies with the relevant harmonized standard.

Choosing the Ideal Structural Material

Most construction projects require the use of hundreds of different materials. These range from concrete of all different specifications, structural steel of different specifications, clay, mortar, ceramics, wood, etc. In terms of a load bearing structural frame, they will generally consist of structural steel, concrete, masonry, and/or wood, using a suitable combination of each to produce an efficient structure. Most commercial and industrial structures are primarily constructed using either structural steel or reinforced concrete. When designing a structure, an engineer must decide which, if not both, material is most suitable for the design. There are many factors considered when choosing a construction material. Cost is commonly the controlling element; however,

other considerations such as weight, strength, constructability, availability, sustainability, and fire resistance will be taken into account before a final decision is made.

Cost—The cost of these construction materials will depend entirely on the geographical location of the project and the availability of the materials. Just as the price of gasoline fluctuates, so do the prices of cement, aggregate, steel, etc. Reinforced concrete derives about half of its construction costs from the required form-work. This refers to the lumber necessary to build the "box" or container in which  the concrete is poured and held until it cures. The expense of the forms makes precast concrete a popular option for designers due to the reduced costs and time. Due to the fact that steel is sold by the pound it is the responsibility of the structural designer to specify the lightest members possible while still maintaining a safe structural design. An additional method of reducing expenditures in design is to use many of the same size steel members as opposed to many unique members.

Strength/weight ratio—Construction materials are commonly categorized by their strength to weight ratio, or specific strength. This is defined as the strength of a material over its density. This gives an engineer an indication as to how useful the material is in comparison to its weight, with the weight being a direct indication of its cost (typically) and ease of construction. Concrete is typically ten times stronger in compression than in tension, giving it a higher strength to weight ratio in compression, only.

Sustainability—Many construction companies and material vendors are making changes to be a more environmentally friendly company. Sustainability has become an entirely new consideration for materials that are to be placed into the environment for generations of time. A sustainable material will be one that has minimal effect on the environment, both at the  time of installation as well as throughout the life cycle of the material. Reinforced concrete and structural steel both have the ability to be a sustainable construction option, if used properly. Over 80% of structural steel members fabricated today come from recycled metals, called A992 steel. This member material is cheaper, as well as having a higher strength to weight ratio, than previously used steel members (A36 grade).

Fire resistance—One of the most dangerous hazards to a building is a fire hazard. This is especially true in dry, windy climates and for structures constructed using wood. Special considerations must be taken into account with structural steel to ensure it is not under a dangerous fire hazard condition. Reinforced concrete characteristically does not pose a threat in the event of fire and even resists the spreading of fire, as well as temperature changes. This makes concrete an excellent insulation, improving the sustainability of the building it surrounds by reducing the required energy to maintain climate.

Corrosion—When choosing a structural material, it is important to consider the life cycle of the building. Some materials are susceptible to corrosion from their surrounding elements, such as water, heat, humidity, or salt. Special considerations must be taken into account during the installation of a structural material to prevent any potential corrosion hazards. This must also be made clear to the occupants of the building because there may or may not be a necessary maintenance requirement to prevent corrosion.

Structural Steel

Characteristics—Structural steel differs from concrete in its attributed compressive strength as well as tensile strength.

Strength—Having high strength, stiffness, toughness, and ductile properties, structural steel is one of the most commonly used materials in commercial and industrial building construction.

Constructability—Structural steel can be developed into nearly any shape, which are either bolted or welded together in construction. Structural steel can be erected as soon as the materials are delivered on site, whereas concrete must be cured at least 1-2 weeks after pouring before construction can continue, making steel a schedule-friendly construction material.

Unit 12　Steel Works

　　Fire resistance—Steel is inherently a noncombustible material. However, when heated to temperatures seen in a fire scenario, the strength and stiffness of the material is significantly reduced. The International Building Code requires steel be enveloped in sufficient fire-resistant materials, increasing overall cost of steel structure buildings.

　　Corrosion—Steel, when in contact with water, can corrode, creating a potentially dangerous structure. Measures must be taken in structural steel construction to prevent any lifetime corrosion. The steel can be painted, providing water resistance. Also, the fire resistance material used to envelope steel is commonly water resistant.

　　Mold—Steel provides a less suitable surface environment for mold to grow than wood.

　　The tallest structures today (commonly called "skyscrapers" or high-rise) are constructed using structural steel due to its constructability as well as its high strength-to-weight ratio. In comparison, concrete, while being less dense than steel, has a much lower strength-to-weight ratio. This is due to the much larger volume required for a structural concrete member to support the same load; steel, though denser, does not require as much material to carry a load. However, this advantage becomes insignificant for low-rise buildings, or those with several stories or less. Low-rise buildings distribute much smaller loads than high-rise structures, making concrete the economical choice. This is especially true for simple structures, such as parking garages, or any building that is a simple, rectilinear shape.

　　Structural steel and reinforced concrete are not always chosen solely because they are the most ideal material for the structure. Companies rely on the ability to turn a profit for any construction project, as do the designers. The price of raw materials (steel, cement, coarse aggregate, fine aggregate, lumber for form-work, etc.) is constantly changing. If a structure could be constructed using either material, the cheapest of the two will likely control. Another significant variable is the location of the project. The closest steel fabrication facility may be much further from the construction site than the nearest concrete supplier. The high cost of energy and transportation

will control the selection of the material as well. All of these costs will be taken into consideration before the conceptual design of a construction project is begun.

Combining Steel and Reinforced Concrete

Structures consisting of both materials utilize the benefits of structural steel and reinforced

concrete. This is already common practice in reinforced concrete in that the steel reinforcement is used to provide steel's tensile strength capacity to a structural concrete member. A commonly seen example would be parking garages. Some parking garages are constructed using structural steel columns and reinforced concrete slabs. The concrete will be poured for the foundational footings, giving the parking garage a surface to be built on. The steel columns will be connected to the slab by bolting and/or welding them to steel studs extruding from the surface of the poured concrete slab. Pre-cast concrete beams may be delivered on site to be installed for the second floor, after which a concrete slab may be poured for the pavement area. This can be done for multiple stories. A parking garage of this type is just one possible example of many structures that may use both reinforced concrete and structural steel.

A structural engineer understands that there are an infinite number of designs that will produce an efficient, safe, and affordable building. It is the engineer's job to work alongside the owner(s), contractor(s), and all other parties involved to produce an ideal product that suits everyone's needs. When choosing the structural materials for their structure, the engineer has many variables to consider, such as the cost, strength/weight ratio, sustainability of the material, constructability, etc.

Fire Resistance

Steel loses strength when heated sufficiently. The critical temperature of a steel member is the temperature at which it cannot safely support its load. Building codes and structural engineering standard practice defines different critical temperatures depending on the structural element type, configuration, orientation, and loading characteristics. The critical temperature is often considered the temperature at which its yield stress has been reduced to 60% of the room temperature yield

stress. In order to determine the fire resistance rating of a steel member, accepted calculations practice can be used, or a fire test can be performed, the critical temperature of which is set by the standard accepted to the Authority Having Jurisdiction, such as a building code. In Japan, this is below 400°C. In China, Europe and North America (e.g., ASTM E-119), this is approximately 1000°F-1300°F (530°C-810°C). The time it takes for the steel element that is being tested to reach the temperature set by the test standard determines the duration of the fire-resistance rating. Heat transfer to the steel can be slowed by the use of fireproofing materials, thus limiting steel temperature. Common fireproofing methods for structural steel include intumescent, endothermic, and plaster coatings as well as drywall, calcium silicate cladding, and mineral wool insulating blankets.

Section B Text Exploration

New Words and Expressions

asymmetrical [ˌæsi'metrikəl]	a.	不对称的，不匀称的
categorize ['kætigəraiz]	v.	分类，把……分类
ceramic [si'ræmik]	n.	陶瓷制品；硅酸盐材料
denote [di'nəut]	v.	表示，表明，指示
endothermic ['endəu'θəːmik]	a.	吸热的
fluctuate ['flʌktjueit]	v.	涨落，起伏，波动，动摇不定，不稳定

intumescent [ˌintjuːˈmesənt]	a.	肿胀的，肿大的，扩大的；沸腾的，（油漆等遇热时）泡沸的
minimal [ˈminiməl]	a.	最小的，最少的，最低限度的
profile [ˈprəufail]	n.	外形，外观，轮廓，侧面，剖面，侧面图
susceptible [səˈseptəbl]	a.	易受影响的
variable [ˈvɛəriəbl]	n.	可变物，易变的事物；变量，变项

be taken into account	被考虑
Charpy impact test	夏比冲击试验
comply with	遵守，服从
consist of	由……组成
cross section	断面，剖面，截面，横切片；断面图，剖面图，截面图
fine grain steel	细粒钢
in comparison to	与……相比
in respect to	关于，就……而言
quenched and tempered steel	调质钢
specific strength	比强度
steel grade	钢种，钢的等级
strength/weight ratio	强度/重量比
structural steel	结构钢，型钢，结构钢材，型钢材
thermomechanically rolled steel	热机械轧制钢
weathering steel	耐腐蚀钢
wide flange	宽凸缘

Proper Names

AISI (American Iron and Steel Institute)	美国钢铁学会

Unit 12　Steel Works

ASTM（American Society for Testing and Materials）International　　美国试验与材料协会国际组织

Construction Products Directive（CPD）　　建筑产品指令

I　True and false.

1. Structural steel shapes, sizes, composition, strengths, storage practices, etc., are not regulated by standards in most industrialized countries. (□T　□F)
2. The concept of CE marking for all construction products and steel products is introduced by the Construction Products Directive (CPD). (□T　□F)
3. Structural steel differs from concrete in its attributed compressive strength as well as tensile strength. (□T　□F)
4. Structural steel and reinforced concrete are not always chosen solely because they are the most ideal material for the structure. (□T　□F)
5. The critical temperature is often considered the temperature at which its yield stress has been reduced to 50% of the room temperature yield stress. (□T　□F)

II　Choose the best answer according to the text.

1. HSS-shape (Hollow structural section also known as SHS) includes the following cross sections except _____.
   A. square　　　　　　　　B. rectangular
   C. diamond　　　　　　　D. circular
2. Typical grades are described as "S275J2" or "S355K2W" and in these examples, "W" denotes _____.
   A. structural steel
   B. the yield strength in newtons per square millimetre or the equivalent megapascals
   C. the materials toughness by reference to Charpy impact test values
   D. weathering steel
3. Which of the following is commonly the controlling element in choosing a

construction material?

A. Weight.   B. Strength.
C. Availability.   D. Cost.

4. Both _____ have the ability to be a sustainable construction option, if used properly.

A. reinforced concrete and structural steel
B. concrete and steel
C. concrete and steel bar
D. structural steel and fireproofing material

5. The tallest structures today (commonly called "skyscrapers" or high-rise) are constructed using structural steel due to its constructability as well as its _____.

A. high strength-to-weight ratio   B. low cost
C. sustainability   D. fire resistance

III Translation.

1. 这些建筑材料的成本完全取决于该项目的地理位置和这些材料的可得性。

    A. The cost of these construction materials will depend partly on the geographical location of the project and the sustainability of the materials.
    B. The cost of these construction materials will depend partly on the quality and the sustainability of the materials.
    C. The cost of these construction materials will depend entirely on the geographical location of the project and the availability of the materials.
    D. The cost of these construction materials will depend entirely on the quality and the transportation of the materials.

2. 有些材料容易受到周边因素如水、热、湿度或盐的腐蚀。

    A. All materials are affected by corrosion from their surrounding elements, such as water, heat, humidity, or salt.
    B. Some materials are susceptible to corrosion from their surrounding elements, such as water, heat, humidity, or salt.
    C. Some materials are susceptible to corrosion from their surrounding elements, such as water, weathering, humidity, or salt.
    D. Some materials are susceptible to environmental conditions, such as water, heat, humidity, or weathering.

3. 结构钢具有高强度、高刚度、高韧性及高延性的性能，是商业建筑和工业建筑最常选用的材料之一。

   A. Having high compressive strength, stiffness, toughness, and fire resistance properties, structural steel is the most commonly used materials in commercial and industrial building construction.

   B. Having high tensile strength, stiffness, toughness, and corrosion resistance properties, structural steel is one of the most commonly used materials in commercial and industrial building construction.

   C. Having high strength, stiffness, toughness, and ductile properties, structural steel is the most commonly used materials in commercial and civil building construction.

   D. Having high strength, stiffness, toughness, and ductile properties, structural steel is one of the most commonly used materials in commercial and industrial building construction.

Section C Supplementary Reading

US Steel Tower

US Steel Tower, also known as the Steel Building (formerly USX Tower) is a 64-story, 256.34 m (841.0 ft) skyscraper with 2,300,000 sq ft (210,000 m²) of leasable space at 600 Grant Street in downtown Pittsburgh, Pennsylvania. It is the tallest skyscraper in Pittsburgh, the fourth tallest building in Pennsylvania, the 41st tallest in the United States, and the 187th tallest building in the world. It held its opening dedication on September 30, 1971.

The tower's original name when completed was the US Steel Building and was changed to USX Tower in 1988. The name was finally changed back to the US Steel Tower in January 2002 to reflect US Steel's new corporate identity (USX was the 1990s combined oil/energy/steel conglomerate). The United States Steel Corporation, more commonly known as US Steel, is an American integrated steel producer with major production operations in the United States, Canada, and Central Europe. Although no longer the owner of the building, US Steel is one of the largest tenants. The building is located at 600 Grant Street, ZIP code 15219.

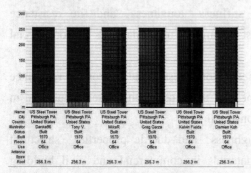

History

In the planning stages, US Steel executives considered making the building the world's tallest, but settled on 840 ft (260 m) and the distinction of being the tallest building outside New York and Chicago. However, it eventually lost even that distinction to newer buildings erected across the United States. Prior to 1970, the tallest building in Pittsburgh, at 44 stories, was the Gulf Building, now known as Gulf Tower. In 2010 Schindler started modernizing the elevators with Schindler Miconic 10 System. In 2014 they started another mod with Schindler Port Technology.

The US Steel Tower is architecturally noted for its triangular shape with indented corners. The building also made history by being the first to use liquid-filled fireproofed columns. US Steel deliberately placed the massive steel columns on the exterior of the building to showcase a new product called Cor-ten steel. Cor-ten resists the corrosive effects of rain, snow, ice, fog, and other meteorological conditions by forming a coating of dark brown oxidation over the metal, which inhibits deeper penetration and doesn't need painting and costly rust-prevention maintenance over the years. The initial weathering of the material resulted in a discoloration of the surrounding city sidewalks, as well as other nearby buildings. A cleanup effort was orchestrated by the corporation once weathering was complete to undo this damage, but the sidewalks still have a decidedly rusty tinge. The Cor-Ten steel for the building was made at the former US Steel Homestead Works. Homestead Steel Works was a large steel works located on the Monongahela River at Homestead, Pennsylvania in the United States. It was developed in the nineteenth century as an extensive plant served by tributary coal and

iron fields, a railway 425 miles long, and a line of lake steamships.

Rockwell International, which had its headquarters in the building, displayed a large model of the Rockwell-designed NASA Space Shuttle in the building's lobby until the company moved its headquarters to Southern California in 1988.

In May 1986 two men parachuted off the roof of the tower before being arrested by Pittsburgh Police near the Civic Arena.

The tower contains over 44,000 US tons (40,000 metric tons) of structural steel, and almost an acre of office space per floor. Currently, the largest tenant of the building is UPMC who occupies 500,000 sq ft (46,000 m^2) of office space within the tower. In 2007 UPMC added signage to the top of the tower for the first time in its history, a move that was criticized by some involved in the construction of the structure.

Features

1. Water filled columns

Although accepted for bridges, etc., exposing structural Cor-Ten in a building was not possible using standard construction methods because steel needed to be protected by concrete or an applied insulation to meet the fire-protection requirements of building codes. A building

code is a set of rules that specify the standards for constructed objects such as buildings and non-building structures. Buildings must conform to the code to obtain planning permission, usually from a local council. The main purpose of building codes is to protect public health, safety and general welfare as they relate to the construction and occupancy of buildings and structures. Fire protection was achieved for the 18 columns of this tower by making them hollow and filled with a water/ antifreeze/rust inhibitor mixture, a technique patented in the 19th century. In another building cited in 1970, the horizontal beams were also hollow and interconnected with the columns, the entire system tested to be leakproof.

2. Internal systems

The US Steel Tower features several redundant systems that have allowed the building to remain free of unplanned service interruptions since it was constructed. It is fed by two redundant water mains, one from Grant Street and one from 7th Avenue. Both are fully maintained and tested annually. There is a fail over system in place, and either main will automatically meet the water demands of the building in the event of a failure. In addition, the building has four redundant water pumps, any one of which can meet the needs of the entire building. The building also has four redundant electrical feeds, which come from several substations. Finally, the building has fully redundant heating and cooling systems, including two boilers and two air chillers. The heating boilers can burn either natural gas or No. 2 fuel oil. There is no fail over, but manual adjustment of the system in the event of a supply shortage takes only minutes.

3. Signage

The University of Pittsburgh Medical Center leased several floors of the tower, which now serves as the institution's headquarters, in 2007. To go along with this lease, the company also purchased new signs reading "UPMC" for the top of each three sides of the building. The Pittsburgh Planning Commission approved the 20-foot (6.1 m) signs, and the majority of the letters were installed via helicopter lift on June 7 – 8, 2008.

4. Nativity scene

The Pittsburgh Crèche is a large-scale crèche, or nativity scene, located on the outside courtyard of the US Steel Tower in downtown Pittsburgh, Pennsylvania. Since 1999, the crèche appears annually during the winter season from November's Light Up Night to Epiphany in January. It is the only authorized replica of the nativity scene in Saint Peter's Basilica in Rome. It is sponsored by the ecumenical Christian Leaders Fellowship.

The number of figures featured in the crèche has increased over time. The original 1999 display featured baby Jesus, Mary, Joseph, the Three Kings, two shepherds, and an assortment of animals. The next year, Simonelli added an angel that hung over the crib and more animals. In 2001, a woman and child were added. In 2002, JE Scenic Technologies added a kneeling shepherd, and later, they created a reclining cow. Two more shepherds and two more angels were later included, bringing the total figures in the display to 20.

5. Roof

Unlike many buildings of similar heights, the US Steel Tower does not taper in width from its lower floors to its higher floors. Accordingly, the tower sports the "largest roof in the world at its height or above", at a size of approximately one acre. This flat expanse was once used as a heliport, but as of January 2012, it had sat dormant for 20 years.

An organization known as the High Point Park Investigation was formed to explore the possibility of converting the dormant roof of the US Steel Tower into an attraction of some sort—a "pinnacle of perspective where people go to see the view, a signature landmark like the Eiffel Tower or the Empire State Building". This transformation could take the form of a nature park, a gallery space, or some other type of attraction. The High Point Park Investigation is based at Carnegie Mellon University's STUDIO for Creative Inquiry and has received the endorsement of regional organizations including the Pittsburgh Parks Conservancy and VisitPittsburgh.com. As of January 2010, the building's owner had expressed no interest in developing the roof of the tower, but public interest in the potential of such a project has been high.

In March 2013 a group of architects and designers finished detailed plans for a glass covered 2 floor atrium at the top of the skyscraper; however the building management has not responded thus far to requests for potential construction.

Sightseeing

On clear days, it is possible to spot the US Steel Tower from as far as 50 miles (80 km) away, from the top of Chestnut Ridge in the Laurel Highlands southeast of the city.

Fictional Portrayals

The US Steel Tower makes an appearance in the movie *Dogma* and figures prominently in *Sudden Death*, *Boys on the Side*, *Striking Distance*, *Jack Reacher* and the video for the 2010 rap song *Black and Yellow*. It can also be seen in the movie *The Dark Knight Rises*. It is the setting for the Syfy network's series of web shorts, *The Mercury Men*. It can also be seen in George A. Romero's 1978 horror film, *Dawn of the Dead*.

The Tower has also been portrayed in multiple video games set in and around Pittsburgh. It can be seen in the *Fallout 3* downloadable mission pack The Pitt. It is also viewable in *The Last of Us* sporting the letters "NA" at the top instead of the usual "UPMC".

Unit 13
Tunneling

Section A Text

Tunnel Engineering

A tunnel is an underground passageway, dug through the surrounding soil, earth, rock and enclosed except for entrance and exit, commonly at each end. A pipeline is not a tunnel, though some recent tunnels have used immersed tube construction techniques rather than traditional tunnel boring methods.

A tunnel may be for foot or vehicular road traffic, for rail traffic, or for a canal. The central portions of a rapid transit network are usually in tunnel. Some tunnels are aqueducts to supply water for consumption or for hydroelectric stations or are sewers. Utility tunnels are used for routing steam, chilled water, electrical power or telecommunication cables, as well as connecting buildings for convenient passage of people and equipment.

A tunnel is relatively long and narrow; the length is often much greater than twice the diameter, although similar shorter excavations can be constructed, such as cross passages between tunnels.

The definition of what constitutes a tunnel can vary widely from source to source. For example, the definition of a road tunnel in the United Kingdom is defined as "a subsurface highway structure enclosed for a length of 150 m (490 ft) or more". In the United States, the NFPA definition of a tunnel is "An underground structure with a design length greater than 23 m (75 ft) and a diameter greater than 1,800 millimeters (5.9 ft)."

In the UK, a pedestrian, cycle or animal tunnel beneath a road or railway is called a subway, while an underground railway system is differently named in different cities, the "Underground" or the "Tube" in London, the "Subway" in Glasgow, and the "Metro" in Newcastle. The place where a road, railway, canal or watercourse passes under a footpath, cycle way, or another road or railway is most commonly called a bridge or, if passing under a canal, an aqueduct. Where it is important to stress that it is passing underneath, it may be called an underpass, though the official term when passing under a railway is an underbridge. A longer underpass containing a road, canal or railway is normally called a "tunnel", whether or not it passes under another item of infrastructure. An underpass of any length under a river is also usually called a "tunnel", whatever mode of transport it is for.

In the US, the term "subway" means an underground rapid transit system, and the term pedestrian underpass is used for a passage beneath a barrier. Rail station platforms may be connected by pedestrian tunnels or footbridges.

Geotechnical Investigation and Design

A major tunnel project must start with a comprehensive investigation of ground conditions by collecting samples from boreholes and by other geophysical techniques. An informed choice can then be made of machinery and methods for excavation and ground support, which will reduce the risk of encountering unforeseen ground conditions. In planning the route, the horizontal and vertical alignments can be selected to make use of the best ground and water conditions. It is common practice to locate a tunnel deeper than otherwise would be required, in order to excavate through solid rock or

other material that is easier to support during construction.

Conventional desk and preliminary site studies may yield insufficient information to assess such factors as the blocky nature of rocks, the exact location of fault zones, or the stand-up times of softer ground. This may be a particular concern in large-diameter tunnels. To give more information, a pilot tunnel (or "drift tunnel") may be driven ahead of the main excavation. This smaller tunnel is less likely to collapse catastrophically should unexpected conditions be met, and it can be incorporated into the final tunnel or used as a backup or emergency escape passage. Alternatively, horizontal boreholes may sometimes be drilled ahead of the advancing tunnel face.

Other Key Geotechnical Factors

"Stand-up time" is the amount of time a newly excavated cavity can support itself without any added structures. Knowing this parameter allows the engineers to determine how far an excavation can proceed before support is needed, which in turn affects the speed, efficiency, and cost of construction. Generally, certain configurations of rock and clay will have the greatest stand-up time, while sand and fine soils will have a much lower stand-up time.

Groundwater control is very important in tunnel construction. Water leaking into a tunnel or vertical shaft will greatly decrease stand-up time, causing the excavation to become unstable and risking collapse. The most common way to control groundwater is to install dewatering pipes into the ground and to simply pump the water out.

A very effective but expensive technology is ground freezing, using pipes which are inserted  into the ground surrounding the excavation, which are then cooled with special refrigerant fluids. This freezes the ground around each pipe until the whole space is surrounded with frozen soil, keeping water out until a permanent structure can be built.

Tunnel cross-sectional shape is also very important in determining stand-up time. If a tunnel excavation is wider than it is high, it will have a harder time supporting itself, decreasing its stand-up time. A square or rectangular excavation is more difficult to make self-supporting, because of a concentration of stress at the corners.

Construction

Tunnels are dug in types of materials varying from soft clay to hard rock. The method of tunnel construction depends on such factors as the ground conditions, the ground water conditions, the length and diameter of the tunnel drive, the depth of the tunnel, the logistics of supporting the tunnel excavation, the final use and shape of the tunnel and appropriate risk management.

There are three basic types of tunnel construction in common use.

(1) Cut-and-cover tunnel, constructed in a shallow trench and then covered over.

(2) Bored tunnel, constructed in situ, without removing the ground above. They are usually of circular or horseshoe cross-section.

(3) Immersed tube tunnel, sunk into a body of water and laid on or buried just under its bed.

Cut-and-cover Tunnel

Cut-and-cover is a simple method of construction for shallow tunnels where a trench is excavated and roofed over with an overhead support system strong enough to carry the load of what is to be built above the tunnel. Two basic forms of cut-and-cover tunneling are available.

(1) Bottom-up method: A trench is excavated, with ground support as necessary, and the tunnel is constructed in it. The tunnel may be of in situ concrete, precast concrete, precast arches, or corrugated steel arches; in early days brickwork was used. The trench is then carefully back-filled and the surface is reinstated.

(2) Top-down method: Side support walls and capping beams are constructed from ground

level by such methods as slurry walling or contiguous bored piling. Then a shallow excavation allows making the tunnel roof of precast beams or in situ concrete. The surface is then reinstated except for access openings. This allows early reinstatement of roadways, services and other surface features. Excavation then takes place under the permanent tunnel roof, and the base slab is constructed.

Shallow tunnels are often of the cut-and-cover type (if under water, of the immersed-tube type), while deep tunnels are excavated, often using a tunneling shield. For intermediate levels, both methods are possible.

Large cut-and-cover boxes are often used for underground metro stations, such as CanaryWharf tube station in London. This construction form generally has two levels, which allows economical arrangements for ticket hall, station platforms, passenger access and emergency egress, ventilation and smoke control, staff rooms, and equipment rooms. The interior of CanaryWharf station has been likened to an underground cathedral, owing to the sheer size of the excavation. This contrasts with many traditional stations on London Underground, where bored tunnels were used for stations and passenger access. Nevertheless, the original parts of the London Underground network, the Metropolitan and District Railways, were constructed using cut-and-cover. These lines pre-dated electric traction and the proximity to the surface was useful to ventilate the inevitable smoke and steam.

A major disadvantage of cut-and-cover is the widespread disruption generated at the surface level during construction. This, and the availability of electric traction, brought about London Underground's switch to bored tunnels at a deeper level towards the end of the 19th century.

Boring Machines

Tunnel boring machines (TBMs) and associated back-up systems are used to highly automate the entire tunneling process, reducing tunneling costs. In certain predominantly urban applications, tunnel boring is viewed as quick and cost effective alternative to laying surface rails and roads. Expensive compulsory purchase of buildings and land, with potentially lengthy planning inquiries, is eliminated. Disadvantages of TBMs arise from their usually large size—the difficulty of transporting the large TBM to the site of tunnel construction, or (alternatively) the high cost of assembling the TBM on-site, often within the confines of the tunnel being constructed.

There are a variety of TBM designs that can operate in a variety of conditions, from hard rock to soft water-bearing ground. Some types of TBMs, the bentonite slurry and earth-pressure balance machines, have pressurized compartments at the front end, allowing them to be used in difficult conditions below the water table. This pressurizes the ground ahead of the TBM cutter head to balance the water pressure. The operators work in normal air pressure behind the pressurized compartment, but may occasionally have to enter that compartment to renew or repair the cutters. This requires special precautions, such as local ground treatment or halting the TBM at a position free from water. Despite these difficulties, TBMs are now preferred over the older method of tunneling in compressed air, with an air lock/decompression chamber some way back from the TBM, which required operators to work in high pressure and go through decompression procedures at the end of their shifts, much like deep-sea divers.

In February 2010, Aker Wirth delivered a TBM to Switzerland, for the expansion of the Linth-Limmern Power Stations located south of Linthal in the canton of Glarus. The borehole has a diameter of 8.03 m (26.3 ft). The four TBMs used for excavating the 57-kilometre (35 mi) Gotthard Base Tunnel, in Switzerland, had a diameter of about 9 m (30 ft). A larger TBM was built to bore the Green Heart Tunnel as part of the HSL-Zuid in the Netherlands, with a diameter of 14.87 m (48.8 ft). This in turn was superseded by the Madrid M30 ringroad, Spain, and the Chong Ming tunnels in Shanghai, China. All of these machines were built at least partly by Herrenknecht. As of August 2013, the world's largest TBM is "Big Bertha", a 57.5-foot (17.5 m) diameter

machine built by Hitachi Zosen Corporation, which is digging the Alaskan Way Viaduct replacement tunnel in Seattle, Washington (US).

Immersed Tube Tunnel

An immersed tube is a kind of underwater tunnel composed of segments, constructed elsewhere and floated to the tunnel site to be sunk into place and then linked together. They are commonly used for road and rail crossings of rivers, estuaries and sea channels/harbours. Immersed tubes are often used in conjunction with other forms of tunnel at their end, such as a cut and cover or bored tunnel, which is usually necessary to continue the tunnel from near the water's edge to the entrance (portal) at the land surface.

The tunnel is made up of separate elements, each prefabricated in a manageable length, then having the ends sealed with bulkheads so they can be floated. At the same time, the corresponding parts of the path of the tunnel are prepared, with a trench on the bottom of the channel being dredged and graded to fine tolerances to support the elements. The next stage is to place the elements into place, each towed to the final location, in most cases requiring some assistance to remain buoyant. Once in position, additional weight is used to sink the element into the final location, this being a critical stage to ensure each piece is aligned correctly. After being put into place the joint between the new element and the tunnel is emptied of water then made water tight, this process continuing sequentially along the tunnel.

The trench is then backfilled and any necessary protection, such as rock armour, added over the top. The ground beside each end tunnel element will often be reinforced, to permit a tunnel boring machine to drill the final links to the portals on land. After these stages the tunnel is complete, and the internal fitout can be carried out.

The segments of the tube may be constructed in one of two methods. In the United States, the preferred method has been to construct steel or cast iron tubes which are then lined with concrete. This allows use of conventional shipbuilding techniques, with the segments being launched after assembly in dry docks. In Europe, reinforced concrete box tube construction has been the standard; the sections are cast in a basin which is then flooded to allow their removal.

Section B Text Exploration

New Words and Expressions

align [ə'lain]	v.	使结盟；匹配；使成一行，排列，排成一行
armour ['ɑːmə]	n.	盔甲，装甲，护面
bentonite ['bentənait]	n.	斑脱土，膨润土，皂土，浆土
borehole ['bɔːhəul]	n.	镗孔，钻孔，炮眼
bulkhead ['bʌkhed]	n.	堤岸墙，（地下通道的）挡墙，挡土墙
buoyant ['bɔiənt]	a.	轻快的；有浮力的，上涨的
chill [tʃil]	v.	冷冻，冷藏；使寒心；使感到冷，变冷
compartment [kəm'pɑːtmənt]	n.	隔间，卧车上的小客房；区划
	v.	分隔，划分
contraband ['kɔntrəˌbænd]	n.	走私，走私货，战时禁运品
	a.	禁运的，非法买卖的
cut-and-cover ['kʌtən'kʌvə]	n.	随挖随填法
diameter [dai'æmitə]	n.	直径
egress ['iːgres]	n.	外出，出口
	v.	使外出，外出
excavation [ˌekskə'veiʃən]	n.	挖掘，发掘

harbour ['hɑːbə]	n.	海港；避难所	
	v.	藏匿；入港停泊；庇护	
immerse [i'məːs]	v.	沉浸，使陷入	
pressurize ['preʃəraiz]	v.	密封；增压，使……加压，使……压入	
reinstate [ˌriːin'steit]	v.	使恢复，使复原	
trench [trentʃ]	n.	沟，沟渠，战壕，堑壕	
	v.	掘沟，挖战壕；侵害	
underbridge ['ʌndəbridʒ]	n.	跨线桥	

base slab	底板
bottom-up method	自底向上法
capping beam	压檐梁，压顶梁
cross-sectional shape	横截面形状
decompression chamber	减压室，降压室
geotechnical investigation	岩土工程勘察
hydroelectric station	水力发电站
precast concrete	预制混凝土
reinforced concrete box	钢筋混凝土箱
situ concrete	现浇混凝土
slurry wall	泥浆墙，槽壁，地下连续壁
stand-up time	自稳时间
top-down method	自顶向下法
tunnel boring machine	隧道掘进机，隧道开挖机

Proper Names

National Fire Protection Association (NFPA)　美国国家消防协会

I True and false.

1. In the United States, the definition of a tunnel is a subsurface highway structure enclosed for a length of 150 meters or more. (□T □F)
2. Configurations of rock and clay have a lower stand-up time than sand and fine soils. (□T □F)
3. There are two basic forms of cut-and-cover tunneling—bottom-up method and top-down method. (□T □F)
4. TBMs are too large to transport, but they are much easier to assemble on-site. (□T □F)
5. In the United States, engineers prefer to construct steel or cast iron tubes which are then lined with concrete. (□T □F)

II Choose the best answer according to the text.

1. What should we do first before a major tunnel project starts?
   A. Make a comprehensive investigation of ground conditions.
   B. Collect samples from boreholes by some geophysical techniques.
   C. Make a choice of machinery and methods for excavation.
   D. Plan the route to make use of the best ground and water conditions.
2. What kind of tunnels are often of the cut-and-cover type?
   A. Wide tunnels.
   B. Shallow tunnels.
   C. Tunnels under water.
   D. Deep tunnels.
3. Which of the following is NOT commonly used by immersed tubes?
   A. Road crossings of rivers.
   B. Rail crossings of rivers.
   C. Estuaries and sea channels.
   D. Shallow and deep tunnels.

Unit 13 Tunneling

4. Which of the following expressions is true about the "stand-up time"?
   A. It is the amount of time a newly excavated cavity can support itself with necessary structures.
   B. It determines how far an excavation can proceed after the support was established.
   C. It affects the speed, efficiency, and construction cost of the project.
   D. Sand and fine soils will have a much greater stand-up time than rock and clay.
5. What is the advantage of the tunnel boring machines?
   A. It is not so expensive so that it can reduce tunneling costs.
   B. The large size can highly automate the entire tunneling process.
   C. The compulsory purchase of buildings and land makes the planning inquiries lengthy.
   D. They can operate in a variety of conditions, from hard rock to soft water-bearing ground.

III Translation.

1. "自稳时间"指的是新开挖的凹槽在没有任何附加结构的情况下能够支撑的时间。
   A. "Stand-up time" is the supporting time of a newly excavated cavity without any other structures.
   B. "Stand-up time" is the supporting time of a newly excavated cavity with no added structures.
   C. "Stand-up time" is the amount of time a newly excavated cavity can support itself without any added structures.
   D. "Stand-up time" is the amount of time that a newly excavated cavity can hold itself with no other structures.
2. 在某些主要的城市应用中,隧道钻孔被认为是一种快捷且划算的替代铺设地面轨道和道路的方法。
   A. In some main urban applications, tunnel boring is thought to be a quick and cost effective alternative method to lay ground rails and roads.
   B. In certain predominantly urban applications, tunnel boring is viewed as quick and cost effective alternative to laying surface rails and roads.
   C. In some main cities, the application of tunnel boring is seen as a quick and cost

effective method to lay surface rails and roads.

   D. In certain urban applications, tunnel boring is a quick and effective way to lay surface rails and roads.

3. 地面冻结技术是指先将管子插入到基坑周围的土地中，然后用特殊的冷冻液将这些管子冷却。

   A. Ground freeze technology is to put pipes into the ground surrounding the excavation, and then cooled the pipes with special refrigerant fluids.

   B. Ground freezing technology is to put pipes into the ground around the excavation, which are then cooled with special refrigerant fluids.

   C. Ground freezing technology is to insert pipes into the ground near the excavation, and then use special refrigerant fluids to cool the pipes.

   D. Ground freezing technology is to insert pipes into the ground surrounding the excavation, which are then cooled with special refrigerant fluids.

4. 在规划路线时，可以选择水平和垂直的对齐方式，以便利用地面和水的最佳条件。

   A. When planning the line, we can select the horizontal and vertical alignments to use the best conditions of ground and water.

   B. When we are planning the line, the horizontal and vertical alignments can be applied to take advantage of best ground and water conditions.

   C. When we are planning the route, the horizontal and vertical alignments can be selected to make use of the best conditions of ground and water.

   D. When making the plan, we can choose the method of horizontal and vertical alignments, and use the best ground and water conditions.

5. 沉管是一种水下隧道，由若干个预置段组成。它们是在其他地方建造的，然后被浮运到隧道现场并沉放到位，最后再连接起来。

   A. An immersed tube is a kind of underwater tunnel composed of some segments, which are constructed elsewhere and then floated to the tunnel site to be sunk into place and finally linked together.

   B. An immersed tube is a kind of underwater tunnel consists of some segments, which are made in other places and then floated to the tunnel site to be sunk into place and finally linked together.

   C. An immersed tube is a kind of underwater tunnel consists of some segments. The segments are constructed elsewhere and transported by floating to the tunnel site to be sunk into place and finally linked together.

D. An immersed tube is a kind of underwater tunnel composed of some segments, which are made in other places and then delivered to the tunnel site by floating to be sunk into place and finally linked together.

Section C Supplementary Reading

Famous Ancient and Modern Tunnels

The history of ancient tunnels and tunneling in the world is reviewed in various sources which include many examples of these structures that were built for different purposes. Some well-known ancient and modern tunnels are briefly introduced below.

The qanat or karez of Persia are water management systems used to provide a reliable supply of water to human settlements or for irrigation in hot, arid and semi-arid climates. The deepest known qanat is in the Iranian city of Gonabad, which after 2700 years, still provides drinking and agricultural water to nearly 40,000 people. Its main well depth is more than 360 m (1,180 ft), and its length is 45 km (28 mi).

The Siloam Tunnel was built before 701 BCE for a reliable supply of water, to withstand siege attacks.

The Eupalinian aqueduct on the island of Samos (North Aegean, Greece) was built in 520 BCE by the ancient Greek engineer Eupalinos of Megara under a contract with the local community. Eupalinos organized the work so that the tunnel was begun from both sides of Mount Kastro. The two teams advanced simultaneously and met in the middle with excellent accuracy, something that was extremely difficult in that time. The aqueduct was of utmost defensive importance, since it ran underground, and it was not easily found by an enemy who could otherwise cut off the water supply to Pythagoreion, the ancient capital of Samos. The tunnel's existence was recorded by Herodotus. The precise location of the tunnel was only re-established in the 19th century by German archaeologists.

One of the first known drainage and sewage networks in form of tunnels was constructed at Persepolis in Iran at the same time as the construction of its foundation in 518 BCE. In most places the network was dug in the sound rock of the mountain and then covered by large

pieces of rock and stone followed by earth and rubbles to level the ground. During investigations and surveys, long sections of similar rock tunnels extending beneath the palace area were traced by Herzfeld and later by Schmidt and their archeological teams.

The Via Flaminia, an important Roman road, penetrated the Furlo pass in the Apennines through a tunnel which emperor Vespasian had ordered built in 76 – 77 CE. A modern road, the SS 3 Flaminia, still uses this tunnel, which had a precursor dating back to the 3rd century BCE; remnants of this earlier tunnel (one of the first road tunnels) are also still visible.

The world's oldest tunnel traversing under a water body is claimed to be the Terelek kaya tüneli under Kızıl River, a little south of the towns of Boyabat and Durağan in Turkey, just downstream from where Kizil River joins its tributary Gökırmak. The tunnel is presently under a narrow part of a lake formed by a dam some kilometers further downstream. Estimated to have been built more than 2000 years ago, possibly by the same civilization that also built the royal tombs in a rock face nearby, it is assumed to have had a defensive purpose.

Sapperton Canal Tunnel on the Thames and Severn Canal in England, dug through hills, which opened in 1789, was 3.5 km (2.2 mi) long and allowed boat transport of coal and other goods. Above it the Sapperton Long Tunnel was constructed which carries the "Golden Valley" railway line between Swindon and Gloucester.

The 1791 Dudley canal tunnel is on the Dudley Canal, in Dudley, England. The tunnel is 1.83 mi (2.9 km) long. Closed in 1962 the tunnel was reopened in 1973. The series of tunnels was extended in 1984 and 1989.

Fritchley Tunnel, constructed in 1793 in Derbyshire by the Butterley Company to transport limestone to its ironworks factory. The tunnel is the world's oldest railway tunnel traversed by

rail wagons using gravity and horse haulage. The railway was converted to steam locomotion in

1813 using a Steam Horse locomotive engineered and built by the Butterley company, however reverted to horses. Steam trains used the tunnel continuously from the 1840s when the railway was converted to a narrow gauge. The line closed in 1933. In the Second World War, the tunnel was used as an air raid shelter. Sealed up in 1977 it was rediscovered in 2013 and inspected. The tunnel was resealed to preserve the construction as it was designated an ancient monument.

The 1794 Butterley canal tunnel is 3,083 yd (2,819 m) in length on the Cromford Canal in Ripley, Derbyshire, England. The tunnel was built simultaneously with the 1773 Fritchley railway tunnel. The tunnel partially collapsed in 1900 splitting the Cromford Canal, and has not been used since. The Friends of Cromford Canal, a group of volunteers, are working at fully restoring the Cromford Canal and the Butterley Tunnel.

The 1796 Stoddart Tunnel in Chapel-en-le-Frith in Derbyshire is reputed to be the oldest rail tunnel in the world. The rail wagons were originally horse-drawn.

A tunnel was created for the first true steam locomotive, from Penydarren to Abercynon. The Penydarren locomotive was built by Richard Trevithick. The locomotive made the historic journey from Penydarren to Abercynon in 1804. Part of this tunnel can still be seen at Pentrebach, MerthyrTydfil, Wales. This is arguably the oldest railway tunnel in the world, dedicated only to self-propelled steam engines on rails.

The Montgomery Bell Tunnel in Tennessee, an 88 m long (289 ft) water diversion tunnel, 4.50 m × 2.45 m high (14.8 ft × 8.0 ft), to power a water wheel, was built by slave labour in 1819, being the first full-scale tunnel in North America.

Bourne's Tunnel, Rainhill, near Liverpool, England. 0.0321 km (105 ft) long. Built in the late 1820s, the exact date is unknown, however probably built in 1828 or 1829. This is the first tunnel in the world constructed under a railway line. The construction of the Liverpool to Manchester Railway ran over a horse-drawn tramway that ran from the Sutton collieries to the Liverpool-Warrington turnpike road. A tunnel was bored under the railway for the tramway. As the railway was being constructed the tunnel was made operational, opening prior to the Liverpool tunnels on the Liverpool to Manchester line. The tunnel was made redundant in 1844 when the tramway was dismantled.

Crown Street Station, Liverpool, England, 1829. Built by George Stephenson, a single track railway tunnel 266 m long (873 ft), was bored from Edge Hill to Crown Street to serve the world's first intercity passenger railway terminus station. The station was abandoned in 1836 being too far from Liverpool city centre, with the area converted for freight use. Closed down in 1972, the tunnel is disused. However it is the oldest passenger rail tunnel running under streets in the world.

The 1829 Wapping Tunnel in Liverpool, England at 2.03 km (1.26 mi) long on a twin track railway, was the first rail tunnel bored under a metropolis. The tunnel's path is from Edge Hill in the east of the city to Wapping Dock in the south end Liverpool docks. The tunnel was used only for freight terminating at the Park Lane goods terminal. Currently disused since 1972, the tunnel was to be a part of the Merseyrail metro network, with work started and abandoned because of costs. The tunnel is in excellent condition and is still being considered for reuse by Merseyrail, maybe with an underground station cut into the tunnel for Liverpool University.

1832, Lime Street Railway station tunnel, Liverpool. A two track rail tunnel, 1.811 km (1.125 mi) long was bored under the metropolis from Edge Hill in the east of the city to Lime Street in Liverpool's city center. The tunnel was in use from 1832 being used to transport building materials to the new Lime St station while under construction. The station and tunnel was opened to passengers in 1836. In the 1880s the tunnel was converted to a deep cutting, open to the atmosphere, being four tracks wide.

Box Tunnel in England, which opened in 1841, was the longest railway tunnel in the world at the time of construction. It was dug by hand, and has a length of 2.9 km (1.8 mi).

The Thames Tunnel, built by Marc Isambard Brunel and his son Isambard Kingdom

Brunel opened in 1843, was the first tunnel (after Terelek) traversing under a water body, and the first to be built using a tunnelling shield. Originally used as a foot-tunnel, the tunnel was converted to a railway tunnel in 1869 and was a part of the East London Line of the London Underground until 2007. It was the oldest section of the network, although not the oldest purpose built rail section. From 2010 the tunnel became a part of the London Overground network.

The 3.34 km (2.08 mi) Victoria Tunnel/Waterloo Tunnel in Liverpool, England, was bored under a metropolis opening in 1848. The tunnel was initially used only for rail freight serving the Waterloo Freight terminal, and later freight and passengers serving the Liverpool ship liner terminal. The tunnel's path is from Edge Hill in the east of the city to the north end Liverpool docks at Waterloo Dock. The tunnel is split into two tunnels with a short open air cutting linking the two. The cutting is where the cable hauled trains from Edge Hill were hitched and unhitched. The two tunnels are effectively one on the same center line and are regarded as one. However, as initially the 2,375 m (1.476 mi) long Victoria section was originally cable-hauled and the shorter 862 m (943 yd) Waterloo section was locomotive hauled, two separate names were given, the short section was named the Waterloo Tunnel. In 1895 the two tunnels were converted to locomotive haulage. Used until 1972, the tunnel is still in excellent condition.

The vertex tunnel of the Semmering railway, the first Alpine tunnel, was opened in 1848 and was 1.431 km (0.889 mi) long. It connected rail traffic between Vienna, the capital of Austro-Hungarian Empire, and Trieste, its port.

The Giovi Rail Tunnel through the Appennini Mounts opened in 1854, linking the capital city of the Kingdom of Sardinia, Turin, to its port, Genoa. The tunnel was 3.25 km (2.02 mi) long.

The oldest underground sections of the London Underground were built using the cut-and-cover method in the 1860s, and opened in January 1863. What are now the Metropolitan, Hammersmith & City and Circle lines were the first to prove the success of a metro or subway system.

On June 18, 1868, the Central Pacific Railroad's 1,659-foot (506 m) Summit Tunnel at Donner Pass in the California Sierra Nevada Mountains was opened permitting the establishment of the commercial mass transportation of passengers and freight over the Sierras for the first time. It remained in daily use until 1993 when the Southern Pacific Railroad closed it and transferred all rail traffic through the 10,322-foot (3,146 m) long tunnel built a mile to the south in 1925.

In 1870, after fourteen years of works, the Fréjus Rail Tunnel was completed between France and Italy, being the second oldest Alpine tunnel, 13.7 km (8.5 mi) long. At that time it was the longest in the world.

The third Alpine tunnel, the Gotthard Rail Tunnel opened in 1882 and was the longest rail tunnel in the world, measuring 15 km (9.3 mi).

The 1882 Col de Tende Road Tunnel, at 3.182 km (1.977 mi) long, was one of the first long road tunnels under a pass, running between France and Italy.

The rail Severn Tunnel was opened in late 1886, at 7.008 km (4.355 mi) long, although only 3.62 km (2.25 mi) of the tunnel is actually under the River Severn. The tunnel replaced the Mersey Railway tunnel's longest under water record, which was held for less than a year.

James Greathead, in constructing the City & South London Railway tunnel beneath the Thames, opened in 1890, brought together three key elements of tunnel construction under water: (1) shield method of excavation; (2) permanent cast iron tunnel lining; (3) construction in a compressed air environment to inhibit water flowing through soft ground material into the tunnel heading.

Built in sections between 1890 and 1939, the section of London Underground's Northern line from Morden to East Finchley via Bank was the longest railway tunnel in the world at 27.8 km (17.3 mi) in length.

In 1906 the fourth Alpine tunnel opened, the Simplon Tunnel, linking Paris to Milan. It is 19.7 km (12.2 mi) long, and was the longest tunnel in the world until 1982.

The 1927 Holland Tunnel was the first underwater tunnel designed for automobiles. The construction required a novel ventilation system.

In 1988 the 53.850 km (33.461 mi) long Seikan Tunnel in Japan was completed under the Tsugaru Strait, linking the islands of Honshu and Hokkaido. It was the longest railway tunnel in the world at that time.

Unit 14
Project Management

Section A Text

Approaches and Process Groups of Project Management

Project management is the discipline of initiating, planning, executing, controlling, and closing the work of a team to achieve specific goals and meet specific success criteria. A project is a temporary endeavor designed to produce a unique product, service or result with a defined beginning and end undertaken to meet unique goals and objectives, typically to bring about beneficial change or added value. The temporary nature of projects stands in contrast with business as usual, which are repetitive, permanent, or semi-permanent functional activities to produce products or services. In practice, the management of these two systems is often quite different, and as such requires the development of distinct technical skills and management strategies.

Approaches

There are a number of approaches for managing project activities including lean, iterative, incremental, and phased approaches.

Regardless of the methodology employed, careful consideration must be given to the overall project objectives, timeline, and cost, as well as the roles and responsibilities of all participants and stakeholders.

1. The traditional approach

A traditional phased approach identifies a sequence of steps to be completed. In the "traditional approach", five developmental components of a project can be distinguished: initiation, planning and design, execution and construction, monitoring and controlling systems and completion or closing.

Many industries use variations of these project stages and it is not uncommon for the stages to be renamed in order to better suit the organization. For example, when working on a brick-and-mortar design and construction, projects will typically progress through stages like pre-planning, conceptual design, schematic design, design development, construction drawings, and construction administration. In software development, this approach is often known as the waterfall model, i.e., one series of tasks after another in linear sequence. In software development many organizations have adapted the Rational Unified Process (RUP) to fit this methodology, although RUP does not require or explicitly recommend this practice. Waterfall development works well for small, well-defined projects, but often fails in larger projects of undefined and ambiguous nature. The Cone of Uncertainty explains some of this as the planning made on the initial phase of the project suffers from a high degree of uncertainty. This becomes especially true as software development is often the realization of a new or novel product. In projects where requirements have not been finalized and can change, requirements management is used to develop an accurate and complete definition of the behavior of software that can serve as the basis for software development. While the terms may differ from industry to industry, the actual stages typically follow common steps to problem solving—defining the problem, weighing options, choosing a path, implementation and evaluation.

2. PRINCE2

PRINCE2 is a structured approach to project management released in 1996 as a generic project management method. It provides a method for managing projects within a clearly defined framework.

PRINCE2 focuses on the definition and delivery of products, in particular their quality requirements. As such, it defines a successful project as being output-oriented (not activity-or task-oriented) through creating an agreed set of products that define the scope of the project and provides the basis for planning and control, that is, how then to coordinate people and activities, how to design and supervise product delivery, and what to do if products and therefore the scope of the project has to be adjusted if it does not develop as planned.

In the method, each process is specified with its key inputs and outputs and with specific goals and activities to be carried out to deliver a project's outcomes as defined by its Business Case. This allows for continuous assessment and adjustment when deviation from the Business Case is required.

3. Critical chain project management

Critical chain project management (CCPM) is a method of planning and managing project execution designed to deal with uncertainties inherent in managing projects, while taking into consideration limited availability of resources needed to execute projects.

CCPM is an application of the theory of constraints (TOC) to projects. The goal is to increase the flow of projects in an organization. Applying the first three of the five focusing steps of TOC, the system constraint for all projects, as well as the resources, are identified. To exploit the constraint, tasks on the critical chain are given priority over all other activities. Finally, projects are planned and managed to ensure that the resources are ready when the critical chain tasks must start, subordinating all other resources to the critical chain.

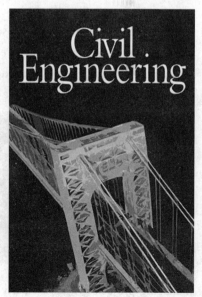

The project plan should typically undergo resource leveling, and the longest sequence of resource-constrained tasks should be identified as the critical chain. In some cases, such as managing contracted sub-projects, it is advisable to use a simplified approach without resource leveling.

4. Process-based management

Process-based management is a management approach that views a business as a collection of processes, managed to achieve a desired result. The processes are managed and improved by organization in purpose of achieving their vision, mission and core value. A clear correlation between processes and the vision supports the company to plan strategies, build a business structure and use sufficient resources that are required to achieve success in the long run.

5. Lean project management

Lean project management uses the principles from lean manufacturing to focus on delivering value with less waste and reduced time. Lean project management has many techniques that can be applied to projects and one of the main methods is standardization.

6. Project Production Management (PPM)

Project Production Management uses principles from operations research, industrial engineering and queuing theory to organize work activities in major capital projects to optimize project delivery.

7. Extreme project management

In critical studies of project management it has been noted that several PERT based models are not well suited for the multi-project company environment of today. Most of them are aimed at very large-scale, one-time, non-routine projects, and currently all kinds of management are expressed in terms of projects.

Using complex models for "projects" spanning a few weeks has been proven to cause

unnecessary costs and low maneuverability in several cases. The generalization of Extreme Programming to other kinds of projects is extreme project management, which may be used in combination with the process modeling and management principles of human interaction management.

8. Benefits realization management

Benefits realization management (BRM) enhances

normal project management techniques through a focus on outcomes (the benefits) of a project rather than products or outputs, and then measuring the degree to which that is happening to keep a project on track. This can help to reduce the risk of a completed project being a failure by delivering agreed upon requirements/outputs but failing to deliver the benefits of those requirements.

In addition, BRM practices aim to ensure the alignment between project outcomes and business strategies. The effectiveness of these practices is supported by recent research evidencing BRM practices influencing project success from a strategic perspective across different countries and industries.

Process Groups

Regardless of the methodology or terminology used, the same basic project management processes or stages of development will be used. Major process groups generally include: initiation, planning, production or execution, monitoring and controlling and closing.

1. Initiating

The initiating processes determine the nature and scope of the project. If this stage is not performed well, it is unlikely that the project will be successful in meeting the business' needs. The key project controls needed here are an understanding of the business environment and making sure that all necessary controls are incorporated into the project. Any deficiencies should be reported and a recommendation should be made to fix them.

The initiating stage should include a plan that encompasses the following areas.

(1) Analyzing the business needs/requirements in measurable goals.

(2) Reviewing of the current operations.

(3) Financial analysis of the costs and benefits including a budget.

(4) Stakeholder analysis, including users, and support personnel for the project.

(5) Project charter, including costs, tasks, deliverables,

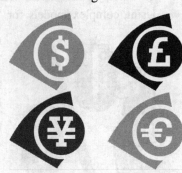

and schedules.

(6) SWOT analysis, including strengths, weaknesses, opportunities, and threats to the business.

2. Planning

After the initiation stage, the project is planned to an appropriate level of detail. The main purpose is to plan time, cost and resources adequately to estimate the work needed and to effectively manage risk during project execution. As with the Initiation process group, a failure to adequately plan greatly reduces the project's chances of successfully accomplishing its goals.

Project planning generally consists of determining how to plan; developing the scope statement; selecting the planning team; identifying deliverables and creating the work breakdown structure; identifying the activities needed to complete those deliverables and networking the activities in their logical sequence; estimating the resource requirements for the activities; estimating time and cost for activities; developing the schedule; developing the budget; risk planning; developing quality assurance measures; gaining formal approval to begin work.

Additional processes, such as planning for communications and for scope management, identifying roles and responsibilities, determining what to purchase for the project and holding a kick-off meeting are also generally advisable.

For new product development projects, conceptual design of the operation of the final product may be performed concurrent with the project planning activities, and may help to inform the planning team when identifying deliverables and planning activities.

3. Executing

While executing we must to know what are the terms we are planned in planning it might be executed synergy. The execution phase ensures that the project management plan's deliverables are executed accordingly. This phase involves proper allocation, coordination and management of human resources and any other resources such as material and budgets. The output of this phase is the project deliverables.

4. Monitoring and controlling

Monitoring and controlling consists of those processes performed to observe project execution so that potential problems can be identified in a timely manner and corrective action can be taken, when necessary, to control the execution of the project. The key benefit is that project performance is observed and measured regularly to identify variances from the project management plan.

Monitoring and controlling includes:

(1) Measuring the ongoing project activities (where we are).

(2) Monitoring the project variables (cost, effort, scope, etc.) against the project management plan and the project performance baseline (where we should be).

(3) Identifying corrective actions to address issues and risks properly (How can we get on track again).

(4) Influencing the factors that could circumvent integrated change control so only approved changes are implemented.

In multi-phase projects, the monitoring and control process also provides feedback between project phases, in order to implement corrective or preventive actions to bring the project into compliance with the project management plan.

Project maintenance is an ongoing process, and it includes continuing support of end-users, correction of errors, updates to the product over time, monitoring and controlling cycle.

In this stage, auditors should pay attention to how effectively and quickly user problems are resolved.

Over the course of any construction project, the work scope may change. Change is a normal and expected part of the construction process. Changes can be the result of necessary design modifications, differing site conditions, material

availability, contractor-requested changes, value engineering and impacts from third parties, to name a few. Beyond executing the change in the field, the change normally needs to be documented to show what was actually constructed. This is referred to as change management. Hence, the owner usually requires a final record to show all changes or, more specifically, any change that modifies the tangible portions of the finished work. The record is made on the contract documents—usually, but not necessarily limited to, the design drawings. The end product of this effort is what the industry terms as-built drawings, or more simply, "as built". The requirement for providing them is a norm in construction contracts. Construction document management is a highly important task undertaken with the aid an online or desktop software system, or maintained through physical documentation. The increasing legality pertaining to the construction industries maintenance of correct documentation has caused the increase in the need for document management systems.

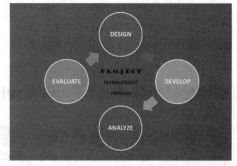

When changes are introduced to the project, the viability of the project has to be re-assessed. It is important not to lose sight of the initial goals and targets of the projects. When the changes accumulate, the forecasted result may not justify the original proposed investment in the project. Successful project management identifies these components, and tracks and monitors progress so as to stay within time and budget frames already outlined at the commencement of the project.

5. Closing

Closing includes the formal acceptance of the project and the ending thereof. Administrative activities include the archiving of the files and documenting lessons learned.

This phase consists of:

Contract closure: Complete and settle each contract and close each contract applicable to the project or project phase.

Project close: Finalize all activities across all of the process groups to formally close the project or a project phase

Also included in this phase is the Post

Implementation Review. This is a vital phase of the project for the project team to learn from experiences and apply to future projects. Normally a Post Implementation Review consists of looking at things that went well and analyzing things that went badly on the project to come up with lessons learned.

Section B Text Exploration

New Words and Expressions

concurrent [kən'kʌrənt]	a.	并发的，一致的，同时发生的
cone [kəun]	n.	圆锥体，圆锥形；球果
	v.	使成锥形
endeavor [in'devə]	n.	努力，尽力
	v.	努力，尽力
explicitly [ik'splisitli]	ad.	明确地，明白地
incremental [ˌinkri'mentəl]	a.	增加的，增值的
inherent [in'hiərənt]	a.	内在的，固有的，天生的
iterative ['itərətiv]	a.	迭代的；重复的，反复的
lean [li:n]	v.	使倾斜，倾斜，倾向；依赖，倚靠
	a.	瘦的，贫乏的
	n.	瘦肉；倾斜，倾斜度
stagger ['stægə]	v.	蹒跚；使交错；使犹豫，犹豫
	n.	蹒跚；交错安排
	a.	交错的，错开的
synergy ['sinədʒi]	n.	协同，协同作用；增效
tangible ['tændʒəbl]	a.	有形的，切实的，可触摸的
	n.	有形资产

Unit 14　Project Management

brick-and-mortar design	实体设计
critical chain project management	关键链项目管理
extreme project management	极限项目管理
lean project management	精益项目管理
linear sequence	线性序列
PERT (program evaluation and review technique)	计划评估和审查技术
PRINCE (project in controlled environment)	受控环境下的项目管理
project and production management	项目与生产管理
rational unified process (RUP)	统一软件过程
resource leveling	资源平衡
SWOT (strengths, weaknesses, opportunities, threats) analysis	态势分析法
theory of constraints (TOC)	约束理论
waterfall development	瀑布式开发

Exercises

I True and false.

1. Waterfall development works well for larger projects of undefined and ambiguous nature. (□T　□F)

2. It is common for the stages of project management to be renamed in order to better suit the organization. (□T　□F)

3. When managing contracted sub-projects, it is suitable to use a simplified approach with resource leveling. (□T　□F)

4. The main principle of lean project management is delivering more value with less waste. (□T　□F)

5. Post Implementation Review is to look at things that went well on the project. (□T　□F)

II Choose the best answer according to the text.

1. What are the five developmental components of a project in the "traditional approach"?
   A. Initiating, planning, executing, controlling, and closing.
   B. Initiation, planning and design, execution and construction, monitoring and controlling systems, completion.
   C. Pre-planning, conceptual design, schematic design, construction drawings, and construction administration.
   D. Defining the problem, weighing options, choosing a path, implementation and evaluation.

2. Critical chain project management is _____ .
   A. a structured approach to project management released in 1996 as a generic project management method
   B. a method of planning and managing project execution designed to deal with uncertainties inherent in managing projects
   C. a management approach that views a business as a collection of processes, managed to achieve a desired result
   D. a method of using the principles from lean manufacturing to focus on delivering value with less waste and reduced time

3. What does benefits realization management focus on?
   A. Outcomes of a project.
   B. Products of a project.
   C. Outputs of a project.
   D. Risk of a product.

4. What processes determine the nature and scope of the project?
   A. Initiating.
   B. Planning.
   C. Executing.
   D. Monitoring and controlling.

5. Which is NOT included in monitoring and controlling?
   A. Measuring the ongoing project activities.
   B. Monitoring the project variables against the project management plan.

C. Identifying corrective actions to address issues and risks properly.

D. Circumventing the influence of the integrated change control.

III Translation.

1. 项目计划通常要经历资源均衡过程，最长的资源受限任务序列应被确定为关键链。

    A. The project plan should typically undergo resource leveling, and the longest sequence of resource-constrained tasks should be identified as the critical chain.

    B. The project plan should usually balance the resource, and the longest sequence of resource-constrained tasks should be confirmed as the critical chain.

    C. The project plan should typically experience resource balance, and the longest sequence of resource-constrained tasks should be identified as the critical chain.

    D. The project plan should typically undergo resource balance, and the longest sequence of resource-limited tasks should be confirmed as the critical chain.

2. 利益实现管理通过注重一个项目的收益而非其产品而提高通常的项目管理技术水平。

    A. Benefits realization management improves normal project management techniques through a focus on benefits of a project rather than products.

    B. Benefits realization management enhances normal project management techniques through a focus on benefits of a project but not products.

    C. Benefits realization management enhances normal project management techniques through a focus on benefits rather than products of a project.

    D. Benefits realization management adopts a focus on benefits but not products of a project to improve normal project management techniques.

3. 不管使用什么方法或术语，项目都将经历相同的基本管理过程或发展阶段。

    A. Regardless of the methodology or terminology used, basic management processes or development stages of the project are the same.

    B. Regardless of the methodology or terminology used, the project will experience the same basic management processes or stages of development.

    C. No matter what methodology or terminology is selected, the same basic project management processes or stages of development will be used.

    D. No matter what methodology or terminology is used, the same basic management processes or stages of development will be applied in a project.

4. 对于新产品开发项目，最终产品运作的概念设计可以与项目计划活动同时进行。
   A. For new product development projects, conceptual design of the final product can start with the project planning activities at the same time.
   B. For new product development projects, conceptual design of the operation of the final product and the project planning activities can begin simultaneously.
   C. For new product development projects, conceptual design of the operation of the final product may be performed with the project planning activities together.
   D. For new product development projects, conceptual design of the operation of the final product may be performed concurrent with the project planning activities.
5. 计划不够充分会大大降低项目成功实现目标的概率。
   A. If the plan is not adequate, the chance of successfully accomplishing goals will decrease greatly.
   B. If the plan is not adequate, it will greatly reduce the chances of successfully accomplishing its goals.
   C. A failure to adequately plan greatly reduces the project's chances of successfully accomplishing its goals.
   D. Inadequate planning will greatly reduce the chances of a successful project being achieved.

Section C Supplementary Reading

Contract Conditions Used for Civil Engineering Work

Standard Conditions of Contract

Over a period of many years there have been a large number of standard forms of conditions of contract introduced. Sometimes these have been developed by particular industries or specialist suppliers, but conditions for more general use have been developed by the major engineering and building institutions, as well as by government and allied organizations. Use of these standard conditions is beneficial because they are familiar to contractors, give greater certainty in operation, and reduce the parties' exposure to risk. Such conditions are often produced by cooperation between contractors' and employers' organizations, with the advice of engineers and other professionals experienced in construction. The documents thus drawn up give a reasonable balance of risk between the parties. However, their clauses are often interdependent, hence any alteration of them must be done with care, and is generally inadvisable because it may introduce uncertainties of interpretation. The main standard conditions used for civil engineering projects are listed below, with an indication of their main provisions.

Contract Conditions Produced by the UK Institution of Civil Engineers

1. ICE conditions of contract for works of civil engineering construction

These are generally known as the ICE conditions and have for many years been the most widely used conditions for UK civil engineering works. They have a long history of satisfactory usage and have been tested in the courts and in arbitration so that the parties to a contract can be confident as to the meaning and interpretation to be placed on these conditions.

The latest edition is the 7th, published in 1999 together with guidance notes, reprinted with amendments in 2003. This edition is known as the Measurement Version to distinguish it from other ICE types of contract based on this established standard.

The principal provisions of the Measurement Version are as follows:

The contractor constructs the works according to the designs and details given in drawings and specifications provided by the employer.

The contractor does not design any major permanent works, but may be required to design special items (such as bearing piles whose choice may depend on the equipment he owns) and building services systems, etc.

An independent engineer, designated "the Engineer" is appointed by the employer to supervise construction, ensure compliance with the contract, authorize variations, and decide payments due; but his decisions can be taken by the employer or contractor to conciliation procedures, adjudication and/or arbitration.

The contractor can claim extra payment and/or extension of time for overcoming unforeseen physical conditions, other than weather, which "could not reasonably have been foreseen by an experienced contractor" and for other delays for which the employer is responsible.

Payment is normally made by re-measurement of work done at rates tendered against items listed in bills of quantifies, which can also include lump sums.

A particular advantage of the ICE conditions is that interpretation of the provisions of the contract lies in the hands of an independent Engineer, who is not a party to the contract, but is required to "act impartially within the terms of the contract having regard to all the circumstances". This gives assurance to both employer and contractor that their interests and obligations under the contract will be fairly dealt with. Also the contractor is paid for overcoming difficulties he could not reasonably have foreseen. Both these matters reduce the contractor's risks, making it possible for him to bid his lowest economic price. This benefits the employer, since the initial price is low and he does not pay out to cover risks which may not occur.

The ICE conditions contain many other provisions that have stood the test of time. These include requirements for early notice of potential delays and problems such as adverse ground conditions and provisions for submission and assessment of claims and valuation of variations. Properly drawn up and administered, a contract under these conditions appears fair to both parties, and the percentage of contracts ending in a dispute which goes to arbitration is very small.

2. ICE conditions for ground investigation

These conditions are based on those described under the ICE conditions but allow for the investigative nature of the work and the need for reports and tests. A schedule of rates may be used instead of a bill of quantifies. The need for a maintenance period and for

Unit 14 Project Management

retention money is left to the drafter, and will depend on whether permanent works, such as measuring devices, are included. The existing (1983) edition is now out of date and a new version is being drafted for issue in 2003 with provisions for dealing with any contaminated land discovered.

3. ICE minor works conditions

These are a shorter and re-written form of the ICE conditions above for use on works which are fully defined at the tender stage and are generally of low value or short duration. The conditions can be successfully used for larger works, but the standard ICE conditions cover many more of the potential problems that can occur on more complex or longer-term projects. Payment arrangements are left open to  be chosen prior to tendering, but are suitable for a single lump sum bid or priced bill of quantities. The 3rd edition of these conditions was published in 2001.

4. ICE design and construct conditions

These conditions were newly produced in 1992 with a 2nd edition in 2001. Known as "the design and construct (D&C) conditions" they follow much of the wording of the Measurement Version but differ significantly from many of the principles of that version. Some of the principal differences are the following:

The employer sets out his required standards and performance objectives for both design 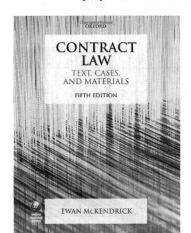 and construction in a document entitled "the Employer's Requirements".

The Contractor develops these requirements and designs and constructs the Works in accordance with them.

The Contractor is responsible for all design matters except any specifically identified in the Contract to be done by others.

An "Employer's Representative" is appointed who supervises the design and construction on behalf of the employer to ensure compliance with the Requirements and that the purpose of the works will be met. He has many duties similar to those of the Engineer under the ICE Measurement Version and is required to behave impartially in regard to certain decisions.

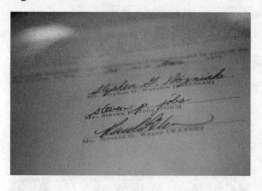

The Employer's Representative can issue instructions to vary the Requirements in reply to which the contractor must submit a quotation for any extra cost or delay in complying with these.

Payment is normally on the basis of a Lump Sum payable in stages, although other means of valuation can be included. However, care is needed if work is re-measured against billed rates, since the contractor could then choose to adopt forms of design that suit the more profitable bill rates he has quoted. Contract conditions used for civil engineering work 43 D&C contracts are often used when the employer's main interest is to have some works built as soon as possible, and he need not, or does not wish to be concerned with the details of the design. The contractor can therefore start construction as soon as he has enough design ready. But where a project offers a wide range of design options, a design and construct contract may not offer an employer the best service because the options chosen by the contractor may tend to be those which suit his plant and workforce best, rather than the interests of the employer. However, if the "Employer's Requirements" are sufficiently extensive and carefully specified, they can go a long way to ensuring coverage of all the employer's needs. It should be the aim of the parties prior to award of contract to arrive at an agreed scheme and specification for the works. Since this form of contract requires extensive input by tenderers their number should be limited to three or four only.

5. ICE: term version

Term contracts have been in use for many years typically to cover repair and maintenance of facilities such as road surfaces or flood defences. This new form, based on the ICE 7th edition and issued in 2002, sets out a background contract which stays in place for a prescribed term of years and under which the Engineer can instruct packages of works to be undertaken as necessary. Works are ordered through a Works Order which defines what is required and supplies any drawings

or specifications not already in the contract. Payment is made by measurement from a schedule of rates in the contract or other agreed means.

The administration and supervision requirements usually follow those of the Measurement Version and will thus be familiar to most engineers. This form can provide a welcome flexibility for employers in procuring irregular items of work or carrying out services which can be called up as and when needed and at short notice. The form may be suitable for some types of framework arrangement.

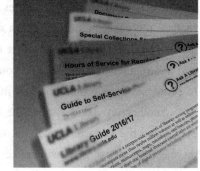

6. ICE engineering and construction contract

This contract was developed from "the New Engineering Contract" (NEC) which was introduced in 1991 and substantially revised in 1993. The NEC is "a family of contracts" comprising versions for construction, sub-contracted works, provision of professional services, and appointment of an adjudicator. The main construction contract was developed and renamed the Engineering and Construction Contract (ECC) which went into a second edition in 1995.

The ECC is formed from "core clauses" which set out the general terms of the contract, "main option clauses" which define valuation and payment methods (one of which must be chosen), and "secondary options clauses" for such as 44 Civil Engineering Project Management bonus, delay damages, and price adjustment. A short form for minor works and a short sub-contract form are also available. The contract requirements are defined in separate sets of data—Works Information and Site Information supplied by the Employer, and Contract Data which set out various pieces of information depending on which options have been chosen.

A project manager appointed by the employer administers the contract on behalf of the employer, assisted by a supervisor on site. A separate adjudicator is appointed to whom the contractor (but not the employer) can take disputes with the project manager or the supervisor for adjudication. But if the employer or the contractor disagrees with the adjudicator's decision either can have the dispute referred to any final tribunal set out in the contract.

The contract attempts to overcome some old problems by several new approaches, but the latter may present some new difficulties. A list of eighteen Compensation Events is prescribed, each of which entitles the contractor to claim extra payment and delay. They include the usual matters of claim such as variation of work, unforeseen conditions, etc. but add unusual weather. The weather data is that supplied by the employer in the Contract Data, and a "weather measurement", could, for instance be rainfall. This definition could give rise to problems of interpretation and may lead to claims even when the weather causes no delay.

Another provision is that the contractor's claims when he experiences a compensation event take the form of quotations which the project manager can accept, return for revision, or reject by advising he will make his own assessment in lieu. Strict time limits of 2 weeks apply to stages of action and response by both contractor and project manager in respect of such quotations and other submissions. These times are tight and may cause difficulties; failure of the project manager to reply within a specified time limit being itself a compensation event.

The stated intent of the drafters of the ECC contract is to stimulate good management. This seems to be achieved by requirements for meetings in a variety of situations, so as to seek advantageous solutions to potential problems, and the tight timetables for responses between the parties.

7. Partnering addendum

This addendum has been issued by the ICE in 2003 to provide for the partners setting down their objectives and any risk sharing provisions formally. The addendum acts as an addition to individual contracts, which may be of any type, and allows for revision as partners leave or are added to a project.